T0275847

LONDON MATHEMATICAL SOCIETY STUDENT TEXTS

Managing editor: Dr C.M. Series, Mathematics Institute
University of Warwick, Coventry CV4 7AL, United Kingdom

LONDON MATHEMATICAL SOCIETY STUDENT TEXTS

Managing editor: Dr C.M. Series, Mathematics Institute
University of Warwick, Coventry CV4 7AL, United Kingdom

London Mathematical Society Student Texts 33

A Primer of Algebraic *D*-modules

S. C. Coutinho
IMPA, Rio de Janeiro

CAMBRIDGE
UNIVERSITY PRESS

Published by the Press Syndicate of the University of Cambridge
The Pitt Building, Trumpington Street, Cambridge CB2 1RP
40 West 20th Street, New York, NY 10011-4211, USA
10 Stamford Road, Oakleigh, Melbourne 3166, Australia

First published 1995

A catalogue record for this book is available from the British Library

Library of Congress cataloging in publication data

Coutinho, S.C.
  A primer of algebraic D-modules / S.C. Coutinho
     p   cm. - (London Mathematical Society Student Texts; 33)
  Includes bibliographical references (p.   -   )
  ISBN 0-521-55119-6 (hardback). 0-521-55908-1 (paperback)
  1. D-modules. I. Title. II. Series.
QA614.3.C68   1995
512'.4-dc20   95-6628 CIP

ISBN   0 521 55119 6 hardback
ISBN   0 521 55908 1 paperback

Transferred to digital printing 2003

For Sérgio Montenegro

CONTENTS

PREFACE

As its title says, this book is only a primer; in particular, you will learn very little 'grammar' from it. That is not surprising; to speak the language of algebraic D-modules fluently you must first learn some algebraic geometry and be familiar with derived categories. Both of these are beyond the bounds of an elementary textbook.

But you can expect to know the answers to two basic questions by the time you finish the book: *what are D-modules?* and *why D-modules?* It is particularly easy to answer the latter, because D-module theory has many interesting applications. Hardly any area of mathematics has been left untouched by this theory. Those that have been touched range from number theory to mathematical physics.

I have tried to include some real applications, but they are not by any means the ones that have caused the greatest impact from the point of view of mathematics at large. To some, they may even seem a little eccentric. That reflects two facts. First, and most important, this is an elementary book. The most interesting applications (to singularity theory and representations of algebraic groups, for example) are way beyond the bounds of such a book. Second, among the applications that were elementary enough to be presented here, I chose the ones that I like the most.

The pre-requisites have been kept to a minimum. So the book should be accessible to final year undergraduates or first year post-graduates. But I have made no effort to write a book that is 'purely algebraic'. Such a book might be possible, but it would not be true. One of the attractions of the theory of D-modules is that it sits across the traditional division of mathematics into algebra, geometry and analysis. It would be a pity to lose that. So this is a book about mathematics, and in it you will find algebras and modules, differential equations and special functions, all in easy conviviality. The introduction contains a detailed description of the pre-requisites.

While writing this book I often worried about the language. I am too fond of English not to shiver at the idea of badly abusing it. But English is not my first language, and I am also well aware of my deficiencies. I can only hope that the many revisions have spared me from sharing the fate of the master of the brigantine in Conrad's *Lord Jim* whose 'flowing English seemed to be derived from a dictionary compiled by a lunatic'.

The material in this book is not original. I have only tried to present the foundations of D-module theory for the beginner that I was, when I started working on the subject ten years ago. In a sense this book is only a compilation: while writing it I have tapped many sources. I have truly plundered the literature for results and exercises to be included in the book. Since it would be very difficult to give references for all of these, I have limited my attributions to the main results and applications.

The book grew out of notes for a basic course on algebraic D-modules taught at the Federal University of Rio de Janeiro. The constant questioning of the students greatly improved the exposition, and directed me to many interesting examples. Many people offered hints, provided references or explained some of the topics to me, particularly G. Meisters, C. Gutierrez, S. Toscano de Melo and A. Pacheco. M.B. Alves and D. Levcovitz read parts of the book and contributed many useful suggestions and corrections. D. Tranah and R. Astley, at Cambridge University Press, put up patiently with the naive questions of a novice and offered many helpful suggestions. Martin Holland read most of the book. His comments saved me from many mistakes while helping to make the exposition clearer; and his unflagging support and friendship kept me at work and made the book possible. To Andrea I owe whatever else was life.

The work in this book was partially supported by a CNPq grant, and benefited from the excellent working conditions generously provided by the Institute of Pure and Applied Mathematics (IMPA).

Rio de Janeiro, April 1995.

INTRODUCTION

This introduction begins with an account of the history of the subject and its relations with other areas of mathematics. This is followed by an overview of the whole book, and a brief description of its chapters.

1. THE WEYL ALGEBRA.

The history of the Weyl algebra begins with the birth of quantum mechanics. The year was 1925. A number of people were trying to develop the principles of the mechanics that was to explain the behaviour of the atom. One of them was Werner Heisenberg. His idea was that this mechanics had to be based on quantities that could actually be observed. In the atomic model of Bohr, there was much talk about orbits; but these are only remotely connected with the things that are actually observed. In the words of Dirac: 'The things that are observed, or which are connected closely with the observed quantities, are all associated with two Bohr orbits and not with just one Bohr orbit: *two* instead of *one*. Now, what is the effect of that?' [Dirac **78**].

To Heisenberg's initial dismay, the 'effect of that' was the introduction of noncommutative quantities. Let Dirac continue with the story:

Suppose we consider all the quantities of a certain kind associated with two orbits, and we want to write them down. The natural way of writing down a set of quantities, each associated with two elements, is in a form like this:

$$
\begin{pmatrix}
\times & \times & \times & \times & \ldots \\
\times & \times & \times & \times & \ldots \\
\times & \times & \times & \times & \ldots \\
\times & \times & \times & \times & \ldots \\
\cdot & \cdot & \cdot & \cdot & \ldots
\end{pmatrix}
$$

an array of quantities, like this, which one sets up in terms of rows and columns. One has the rows connected with one of the states, the columns connected with the other. Mathematicians call a set of quantities like this a matrix.

The sequence of events is a little more intricate than Dirac's comments will have us believe. In 1925 matrices were not part of the toolkit of every physicist, as they now are. What Heisenberg originally introduced were quantum theoretical analogues of the classical Fourier series. These were supposed to describe the dynamical variables of the atomic systems; and they did not commute. Unable to go as far as he had hoped, Heisenberg decided to sum up his ideas in a paper that he presented to Max Born. Born was acquainted with matrices and was the first to realize that matrix theory offered the correct formalism for Heisenberg's ideas.

According to Born's approach, the dynamical variables (velocity, position, momentum) should be represented by matrices in quantum theory. Denoting the position matrix by q and the momentum matrix by p, one may write the equation for a system with one degree of freedom in the form

$$pq - qp = \imath\hbar.$$

The work on matrix mechanics began with Born and his assistant P. Jordan, soon to be joined by Heisenberg himself. Early on they understood that the fundamental equation above could not be realized by finite matrices; the matrices of quantum theory had to be infinite.

Matrix mechanics was soon followed by other formalisms. First came E. Schrödinger's wave mechanics. In this approach everything begins with a partial differential equation; a more familiar object to physicists. There was also Dirac's formalism. He chose the relations among the dynamical variables as his starting point. The dynamical variables subject to those relations were the elements of what he called *quantum algebra*.

From Dirac's point of view, one is interested in polynomial expressions in the dynamical variables momentum, denoted by p, and position, denoted by q. It is assumed that the variables satisfy the (normalized) relation $pq - qp = 1$. This is what we now call the first Weyl algebra. In particular, he showed how one could use the relation between p and q to differentiate polynomial expressions with respect to p and to q. The Weyl algebras of higher index appear when one considers systems with several degrees of freedom. The relations among the operators in this case had been established as early as September 1925 by Heisenberg. Dirac's point of view received a masterly presentation in H. Weyl's pioneering book *The theory of groups and quantum mechanics*, [Weyl **50**].

From that point onwards the mathematicians were ready to take over. Of very special interest is the paper [Littlewood **33**] of D. Littlewood. He begins by saying that although finite dimensional algebras had been intensively studied, the same was not true of algebras of infinite dimensions. One of the algebras he studies is Dirac's quantum algebra. Littlewood carefully constructs (infinite) matrices p and q for which the equation $pq - qp = 1$ is satisfied; see Exercise 1.4.10.

In his paper Littlewood established many of the basic properties of the Weyl algebra. He showed that the elements of the algebra have a canonical form (Ch. 1 ,§2) and that the algebra is a domain (Ch. 2, §1). He also showed that the relation $pq - qp = 1$ is not compatible with any other relation. Or, as we would now say, the only proper ideal of this algebra is zero (Ch. 2, §2).

The modern age in the theory of the Weyl algebra arrived when its connections with Lie algebras were realized. Suppose that **n** is a nilpotent Lie algebra over $\mathbb{C}$. Let $U(\mathbf{n})$ be its enveloping algebra. The quotient of $U(\mathbf{n})$ by a *primitive* ideal is always isomorphic to the Weyl algebra; see [Dixmier **74**, Théorème 4.7.9]. In [Dixmier **63**], the notation A_n was introduced for the algebra that corresponds to the physicist's system with n degrees of freedom. The name Weyl algebra was used by Dixmier as the title of [Dixmier **68**] following, as he says, a suggestion of I. Segal in [Segal **68**].

The increasing interest in noncommutative noetherian rings that followed A. Goldie's work and the intense development of the theory of enveloping algebras of Lie algebras contributed to keeping up the interest in the Weyl algebra. That is not the end of the story though: another theme was added in the seventies under the guise of D-module theory.

2. ALGEBRAIC D-MODULES.

Under the cryptic name of D-*module* hides a modest module over a ring of differential operators. The importance of the theory lies in its manifold applications, which span a vast territory. The representation theory of Lie algebras, differential equations, mathematical physics, singularity theory and even number theory have been influenced by D-modules.

One of the roots of the theory is the idea of considering a differential equation as a module over a ring of differential operators discussed in Ch.

6. This approach goes back at least to the sixties, when it was applied by B. Malgrange to equations with constant coefficients. A turning point was reached in 1971 with M. Kashiwara's thesis 'Algebraic study of systems of partial differential equations', where the same approach was systematically applied to equations with analytic coefficients. In this context, the theory is often called by the alternative name of *algebraic analysis*.

At the same time, in the Soviet Union, I. N. Bernstein was developing the theory of modules over the Weyl algebra. His starting point, however, was entirely different. I.M. Gelfand had asked in the International Congress of Mathematicians of 1954 whether a certain function of a complex variable, which was known to be analytic in the half plane $\Re(z) > 0$, could be extended to a meromorphic function defined in the whole complex plane. The problem remained open until 1968, when M.F. Atiyah and, independently, Bernstein and I.S. Gelfand gave affirmative answers. Both proofs made use of Hironaka's *resolution of singularities*, a very deep and difficult result.

Four years later, Bernstein discovered a new proof of the same result that was very elementary. The key to the proof was a clever use of the Weyl algebra. In his papers, Bernstein introduced many of the concepts that we will study in this book.

Of course the theory of D-modules is not restricted to the Weyl algebra. The theory has two branches: an analytic and an algebraic one; depending on whether the base variety is analytic or algebraic. Highly sophisticated machinery is required in the study of general D-modules, and the most important results cannot be introduced without derived categories and sheaves.

Perhaps the most spectacular result of the theory to date is the *Riemann-Hilbert correspondence* obtained independently by M. Kashiwara and Z. Mebkhout in 1984. This is a result of noble parentage; it can be traced to Riemann's memoir on the hypergeometric function. Its genealogical tree includes the work of Fuchs on differential equations with regular singular points and Hilbert's 21st problem of 1900. Very roughly speaking, the correspondence establishes an (anti-)equivalence between certain differential equations (described in terms of D-modules) and their solution spaces. Unfortunately, the correspondence requires deep results of category theory and cannot be included in an elementary book.

3. THE BOOK: AN OVERVIEW.

This book is about modules over the Weyl algebra, and the point of view is that of algebraic D-modules. It is also fair to say that this is a book about certain aspects of the representation theory of the Weyl algebra. So we will have a lot to say about irreducible modules and the dimension of modules, for example.

The book can be divided into two parts. The first part, which goes up to Ch. 11, is concerned with invariants of modules over the Weyl algebra; the most important being the dimension and the multiplicity.

The first two chapters deal with ring theoretic properties of the Weyl algebra itself. Most importantly, we show that it is a simple domain. Ch. 3 establishes the Weyl algebra as a member of the family of rings of differential operators. There is much talk of derivations in this chapter, and they are put to good use in the next chapter. Ch. 4 contains the first of our applications. It consists of using a *conjecture* about the Weyl algebra to derive the *Jacobian conjecture*.

With Ch. 5 we get into the realm of representation theory. The purpose of the chapter is to describe a few important examples of modules over the Weyl algebra. The relation of these modules with differential equations is found in Ch. 6, which also includes an elementary description of the module of microfunctions. These are used as generalized solutions of differential equations.

Chs. 7 to 9 form a sequence which culminates in the definition of the dimension and multiplicity of a module over the Weyl algebra and the study of their properties. Place of honour is given to Bernstein's inequality: the dimension of a module over the n-th Weyl algebra is an integer between n and $2n$. The modules of minimal dimension form such a nice category that they have a special name: holonomic modules. The whole of Ch. 10 is dedicated to them. Ch. 11 requires a smattering of algebraic geometry; it deals with the relation between D-modules and symplectic geometry. The key concept is another invariant of a D-module, the *characteristic variety*. This allows us to give a geometrical interpretation to the dimension previously defined.

The emphasis now shifts from invariants to operations, which are the theme of the second part. These operations are geometrical constructions which use a polynomial map to produce new D-modules. Since their definition depends on the use of the tensor product, which may not be familiar to some readers, we have included a discussion of them in Ch. 12.

Chs. 13 to 16 contain the definitions and examples of the three main operations that we apply to modules over the Weyl algebra: external products, inverse images and direct images. The results are served in homœopathic doses, since the calculations tend to be crammed with detail. Kashiwara's theorem is one of the pearls of the theory. A simple, rather meagre version is described in Ch. 17, but it will return in greater splendour in Ch. 18. In this chapter we also show that all the operations previously described map holonomic modules to holonomic modules. This is very mystifying since the dimension of a module is *not* preserved by some of these operations.

Finally we return to applications in Chs. 19 and 20. The former is concerned with the global stability of ordinary differential equations. We will discuss conditions under which a system of differential equations with polynomial coefficients has a global stability point. The key lemma has a very neat D-module theoretic proof due to van den Essen. It is discussed in detail in §2. The proof of the stability theorem is sketched in §3. Ch. 20 is about the work of Zeilberger and his collaborators on the automatic computation of sums and integrals. That is, how can D-modules help a computer to calculate a definite integral?

4. PRE-REQUISITES.

All the algebra required in the book can be found in [Cohn **84**]. Like all general rules, this one has exceptions; these are Ch. 11 and Ch. 18, §3. We have already mentioned that one needs to know some algebraic geometry to follow Ch. 11. All the required results will be found in [Hartshorne **77**, Ch. 1]. In §3 of Ch. 18 we rewrite the results of Chs. 14 to 17 using categories. The subjects of these sections are not used anywhere else in the book, except in some exercises.

The book also requires a good knowledge of analysis to be properly appreciated. This is particularly true of examples and applications. However, most of the results we use are part of the standard undergraduate curriculum; and we will give several references to those which are not so well-known. Ch. 19, in particular, requires some results about ordinary differential equations. A good reference for our needs is [Arnold **81**].

The book is linearly ordered and the order is almost, though not quite, total. There are two obvious exceptions: Ch. 11 and §3 of Ch. 18 depend on everything that comes before them, but are not used anywhere else in the book. The only other exceptions are Chs. 4 and 6. We will need Ch. 4, §1 later on, but that will be only in Ch. 14.

CHAPTER 1
THE WEYL ALGEBRA

We will describe the main protagonist of this book, the Weyl algebra, in two different ways: as a ring of operators and in terms of generators and relations.

1. DEFINITION

In this section the Weyl algebra is introduced as a ring of operators on a vector space of infinite dimension. Let us fix some notation. Throughout this book, K denotes a field of characteristic zero and $K[X]$ the ring of polynomials $K[x_1, \ldots, x_n]$ in n commuting indeterminates over K.

The ring $K[X]$ is a vector space of infinite dimension over K. Its algebra of linear operators is denoted by $End_K(K[X])$. Recall that the algebra operations in the endomorphism ring are the addition and composition of operators. The Weyl algebra will be defined as a subalgebra of $End_K(K[X])$.

Let $\hat{x}_1, \ldots, \hat{x}_n$ be the operators of $K[X]$ which are defined on a polynomial $f \in K[X]$ by the formulae $\hat{x}_i(f) = x_i \cdot f$. Similarly, $\partial_1, \ldots, \partial_n$ are the operators defined by $\partial_i(f) = \partial f / \partial x_i$. These are linear operators of $K[X]$. The *n-th Weyl algebra* A_n is the K-subalgebra of $End_K(K[X])$ generated by the operators $\hat{x}_1, \ldots, \hat{x}_n$ and $\partial_1, \ldots, \partial_n$. For the sake of consistency, we write $A_0 = K$.

Note that for $n \geq m$, the action of the operators of A_m on $K[X]$ is well-defined. Thus A_m is a subalgebra of A_n in a natural way. We sometimes write $A_n(K)$ instead of A_n, if it is necessary to make explicit the field over which the algebra is defined.

According to our definition, the elements of A_n are linear combinations over K of monomials in the generators $\hat{x}_1, \ldots, \hat{x}_n, \partial_1, \ldots, \partial_n$. However, one has to be careful when representing the elements of A_n because this algebra is not commutative. This is easily checked, as follows. Consider the operator $\partial_i \cdot \hat{x}_i$ and apply it to a polynomial $f \in K[X]$. Using the rule for the differentiation of a product, we get $\partial_i \cdot \hat{x}_i(f) = x_i \partial f / \partial x_i + f$. In other words,

$$\partial_i \cdot \hat{x}_i = \hat{x}_i \cdot \partial_i + 1$$

where 1 stands for the identity operator. It is better to rewrite this formula using commutators. If $P, Q \in A_n$ then their *commutator* is the operator $[P, Q] = P \cdot Q - Q \cdot P$. The formula above becomes $[\partial_i, \hat{x}_i] = 1$. Similar calculations allow us to obtain formulae for the commutators of the other generators of A_n. These are summed up below:

$$[\partial_i, \hat{x}_j] = \delta_{ij} \cdot 1,$$

$$[\partial_i, \partial_j] = [\hat{x}_i, \hat{x}_j] = 0,$$

where $1 \leq i, j \leq n$. Recall that δ_{ij} is the Kronecker delta symbol: it equals 1 if $i = j$ and zero otherwise. A final observation. We have denoted the operator 'multiplication by x_i' by the symbol $\hat{x}_i$. From now on, we shall follow the standard convention and write x_i for both the variable and the corresponding operator. This tends to make the notation less cluttered. For the same reason we shall dispense with the subscripts for the generators of A_1, and write them simply as x and ∂.

2. CANONICAL FORM

In this section we construct a basis for the Weyl algebra as a K-vector space. This basis is known as the *canonical basis*. If an element of A_n is written as a linear combination of this basis then we say that it is in *canonical form*. Of course, to compare two elements in canonical form it is enough to compare the coefficients of their linear combinations, and that is easily done.

It is easier to describe the canonical basis if we use a multi-index notation. A *multi-index* α is an element of $\mathbb{N}^n$; say $\alpha = (\alpha_1, \ldots, \alpha_n)$. Now by x^α we mean the monomial $x_1^{\alpha_1} \ldots x_n^{\alpha_n}$. The *degree* of this monomial is the *length* $|\alpha|$ of the multi-index α, namely $|\alpha| = \alpha_1 + \ldots + \alpha_n$. Notice that a pair (α, β) of multi-indices in $\mathbb{N}^n$ is itself a multi-index in $\mathbb{N}^{2n}$, so it makes sense to talk of its length.

2.1 PROPOSITION. *The set* $\mathbf{B} = \{x^\alpha \partial^\beta : \alpha, \beta \in \mathbb{N}^n\}$ *is a basis of* A_n *as a vector space over* K.

The proof of this proposition uses a formula for the derivative of polynomials in terms of multi-indices that we state below. The factorial of a multi-index $\beta \in \mathbf{N}^n$ is defined by $\beta! = \beta_1! \ldots \beta_n!$. The formula is written in terms of powers of the operators ∂_i. The proof is left to the reader.

2.2 LEMMA. *Let $\sigma, \beta \in \mathbf{N}^n$ and assume that $|\sigma| \leq |\beta|$. Then $\partial^\beta(x^\sigma) = \beta!$ if $\sigma = \beta$, and zero otherwise.*

PROOF OF THE PROPOSITION: It is easy to see that the elements of **B** generate the Weyl algebra as a vector space. Consider a monomial on the generators of A_n. Using the relations of §1, one shows that if $f \in K[x]$, then $\partial_i \cdot f - f \cdot \partial_i = \partial f / \partial x_i$. That allows us to bring all powers of x's to the left of all the ∂'s. By doing that, the monomial automatically ends up written as a linear combination of the elements of **B**.

Now to the uniqueness. Consider a finite linear combination of elements of **B**, say $D = \sum c_{\alpha\beta} x^\alpha \partial^\beta$. We must show that if some $c_{\alpha\beta}$ is non-zero then $D \neq 0$. But D is a linear operator of $K[X]$. Hence $D \neq 0$ if and only if there exists a polynomial f for which $D(f) \neq 0$. We construct such an f.

Let σ be a multi-index which satisfies $c_{\alpha\sigma} \neq 0$ for some index α, but $c_{\alpha\beta} = 0$, for all indices β such that $|\beta| < |\sigma|$. A straightforward calculation using Lemma 2.2 shows that $D(x^\sigma) = \sigma! \sum_\alpha c_{\alpha\sigma} x^\alpha$. This is non-zero since at least one of the coefficients $c_{\alpha\sigma}$ is non-zero by the choice of σ. Thus $f = x^\sigma$ is the required polynomial.

3. GENERATORS AND RELATIONS

Another way to define the Weyl algebra is by generators and relations. More precisely, we may write the Weyl algebra as a quotient of a free algebra in $2n$ generators. The ideal that we factor out is generated by the relations calculated in §1.

The *free algebra* $K\{z_1, \ldots, z_{2n}\}$ in $2n$ generators is the set of all finite linear combinations of words in $z_1, \ldots, z_{2n}$. Multiplication of two monomials is simple juxtaposition. We may define a surjective homomorphism $\phi : K\{z_1, \ldots, z_{2n}\} \longrightarrow A_n$ by $\phi(z_i) = x_i$ and $\phi(z_{i+n}) = \partial_i$, for $i = 1, 2, \ldots, n$.

Let J be the two-sided ideal of $K\{z_1, \ldots, z_{2n}\}$ generated by $[z_{i+n}, z_i] - 1$ for $i = 1, 2, \ldots, n$ and $[z_i, z_j]$ for $j \neq i + n$ and $1 \leq i, j \leq 2n$. It follows from the relations of §1 that $J \subseteq \ker \phi$. Thus ϕ induces a homomorphism of K-algebras $\hat{\phi} : K\{z_1, \ldots, z_{2n}\}/J \longrightarrow A_n$.

3.1 THEOREM. $\hat{\phi}$ *is an isomorphism.*

PROOF: Exactly as in the proof of Proposition 2.1, we may use the relations between the classes $z_i + J$ to show that every element of $K\{z_1, \ldots, z_{2n}\}/J$ may be written as a linear combination of monomials of the form

$$z_1^{m_1} \ldots z_{2n}^{m_{2n}} + J.$$

By Proposition 2.1, the images of these monomials under $\hat{\phi}$ form a basis of A_n as a vector space over K. In particular, the monomials must be linearly independent in $K\{z_1, \ldots, z_{2n}\}/J$. Hence $\hat{\phi}$ is an isomorphism of vector spaces and, *a fortiori*, an isomorphism of rings.

We may apply this theorem to the construction of automorphisms of A_n. The corollary contains a very important example, which will be used in later chapters.

3.2 COROLLARY. *Let $m < n$ be positive integers. Choose polynomials $f_i \in K[X]$, for $1 \leq i \leq n$, as follows: if $i \leq m$, then f_i is a polynomial in the variables $x_{m+1}, \ldots, x_n$; otherwise $f_i = 0$. The map $\sigma : A_n \longrightarrow A_n$ defined by the formulae*

$$\sigma(x_i) = x_i + f_i,$$

$$\sigma(\partial_i) = \partial_i - \sum_1^n \frac{\partial f_k}{\partial x_i} \partial_k$$

is an automorphism of A_n.

PROOF: Define a homomorphism ϕ of $K\{z_1, \ldots, z_{2n}\}$ to A_n by $\phi(z_i) = x_i + f_i$ and

$$\phi(z_{i+n}) = \partial_i - \sum_1^n \frac{\partial f_k}{\partial . x_i} \partial_k$$

Choose i, j such that $1 \leq i, j \leq n$. It is clear that $\phi([z_i, z_j]) = 0$. Let us calculate $\phi([z_{i+n}, z_j])$. Since ϕ is a ring homomorphism, this is the same as $[\phi(z_{i+n}), \phi(z_j)]$, which equals

$$[\partial_i, x_j + f_j] - [\sum_1^n \frac{\partial f_k}{\partial x_i} \partial_k, x_j + f_j].$$

The first commutator is $\delta_{ij} + \partial f_j/\partial x_i$; the second equals

$$\frac{\partial f_j}{\partial x_i} + \sum_{1}^{n} \frac{\partial f_k}{\partial x_i} \cdot \frac{\partial f_j}{\partial x_k}.$$

Subtracting these two terms, and using the hypotheses on the f's, we get that $\phi([z_{i+n}, z_j]) = \delta_{ij}$. A similar calculation shows that $\phi([z_{i+n}, z_{j+n}]) = 0$. Thus ϕ induces an endomorphism $\sigma = \hat{\phi}$ of A_n.

A similar argument shows that the map τ defined by the formulae

$$\tau(x_i) = x_i - f_i,$$

$$\tau(\partial_i) = \partial_i + \sum_{1}^{n} \frac{\partial f_k}{\partial x_i} \partial_k$$

is an endomorphism of A_n. It is now easy to check that τ is the inverse of σ. Thus σ is an automorphism of A_n.

The automorphisms of A_n play a very important rôle in the theory, as we shall see. In particular, it is not known when an endomorphism of A_n is an automorphism. This question is related to the Jacobian conjecture, and will be discussed in Ch. 4, §4.

4. EXERCISES

4.1 Let $f \in K[X]$. Prove that $[\partial_i, f] = \partial f/\partial x_i$ in A_n.

4.2 Find the canonical form of the elements of A_3 listed below.

(1) $\partial_1^2 \cdot x_1^2$

(2) $\partial_2^3 \cdot x_1 \cdot \partial_3 \cdot x_3 + x_3 \cdot \partial_1 \cdot x_1$

(3) $\partial_1 \cdot x_1 \cdot \partial_2 \cdot \partial_3 \cdot x_3 \cdot x_2$

(4) $\partial_1^3 \cdot x_1^2 + \partial_2^2 \cdot x_2^3$

4.3 Let R be a K-algebra and $\lambda_1, \lambda_2 \in K$. Show that the commutator of two elements of R satisfies $[a, \lambda_1 b_1 + \lambda_2 b_2] = \lambda_1[a, b_1] + \lambda_2[a, b_2]$.

4.4 Let V be an infinite dimensional real vector space with basis $\{u_n : n \in \mathbb{N}\}$. Define $\xi, \eta \in End_{\mathbb{R}} V$ by $\xi(u_n) = u_{n+1}(n+1)^{1/2}$ for all n, $\eta(u_0) = 0$ and

$\eta(u_n) = u_{n-1}n^{1/2}$ for $n > 0$. Let R be the algebra generated by ξ and η. Show that $R \cong A_1(\mathbb{R})$.

4.5 Let R be a ring. Let $a, b, c \in R$. Show that Jacobi's identity $[a, [b, c]] + [c, [a, b]] + [b, [c, a]] = 0$ holds in R.

4.6 Show that if $d \in A_n$ and $f, g \in K[X]$, then $[[d, f], g] = [[d, g], f]$. Hint: Jacobi's identity.

4.7 Let $SL(2, K)$ be the group of 2×2 matrices with coefficients in K and determinant 1. Let $(a_{ij}) \in SL(2, K)$. Show that there exists an automorphism σ of A_1 which satisfies $\sigma(\partial) = a_{11}\partial + a_{12}x$ and $\sigma(x) = a_{21}\partial + a_{22}x$.

4.8 Show that there exists an automorphism $\mathcal{F} : A_n \longrightarrow A_n$ which satisfies $\mathcal{F}(x_i) = \partial_i$ and $\mathcal{F}(\partial_i) = -x_i$, for $i = 1, 2, \ldots, n$.

4.9 Let $M_n(K)$ denote the algebra of $n \times n$ matrices with entries in K. If $A \in M_n(K)$ show that the map $\phi : M_n(K) \rightarrow M_{n+1}(K)$ defined by

$$\phi(A) = \begin{pmatrix} A & 0 \\ 0 & 0 \end{pmatrix}$$

is an injective K-algebra homomorphism. So we may assume that $M_n(K) \subseteq M_{n+1}(K)$. Show that $M_\infty(K) = \bigcup_{n \geq 1} M_n(K)$ is a K-algebra of infinite dimension.

4.10 Consider the following two matrices in $M_\infty(K)$:

$$P = \begin{pmatrix} 0 & 1 & 0 & 0 & \cdots \\ 0 & 0 & 2 & 0 & \cdots \\ 0 & 0 & 0 & 3 & \cdots \\ \vdots & \vdots & \vdots & \vdots & \ddots \end{pmatrix} \quad \text{and } Q = \begin{pmatrix} 0 & 0 & 0 & 0 & \cdots \\ 1 & 0 & 0 & 0 & \cdots \\ 0 & 1 & 0 & 0 & \cdots \\ \vdots & \vdots & \vdots & \vdots & \ddots \end{pmatrix}.$$

Show that the K-algebra of $M_\infty(K)$ generated by P and Q is isomorphic to $A_1(K)$.

CHAPTER 2

IDEAL STRUCTURE OF THE WEYL ALGEBRA

It is time to discuss the ideal structure of the Weyl algebra. We will introduce the degree of an operator and use it to show that the Weyl algebra is a domain whose only proper two-sided ideal is zero.

1. THE DEGREE OF AN OPERATOR.

The degree of an operator of A_n, to be introduced in this section, behaves, in many ways, like the degree of a polynomial. The differences are accounted for by the noncommutativity of A_n.

Let $D \in A_n$. The *degree* of D is the largest length of the multi-indices $(\alpha, \beta) \in \mathbb{N}^n \times \mathbb{N}^n$ for which $x^\alpha \partial^\beta$ appears with non-zero coefficient in the canonical form of D. It is denoted by $deg(D)$. As with the degree of a polynomial, we use the convention that the zero polynomial has degree $-\infty$. An example will suffice: the degree of $2x_1 \partial_2 + x_1 x_2^3 \partial_1 \partial_2$ is 6.

If $D, D' \in A_n$ are written in canonical form, then so is $D + D'$, and one concludes that $deg(D + D') \leq \max\{deg(D), deg(D')\}$. Note that if $deg(D) \neq deg(D')$ then we have equality in the above formula. The formula $deg(DD') = deg(D) + deg(D')$ also holds, but its proof is harder, because A_n is noncommutative. From now on we shall denote by e_i the multi-index all of whose entries are zero, except the i-th entry, which is 1.

1.1 THEOREM. *The degree satisfies the following properties; for $D, D' \in A_n$:*

(1) $deg(D + D') \leq \max\{deg(D), deg(D')\}$.

(2) $deg(DD') = deg(D) + deg(D')$.

(3) $deg[D, D'] \leq deg(D) + deg(D') - 2$.

PROOF: (1) has already been proved. We prove (2) and (3) at the same time, by induction on $deg(D) + deg(D')$. If either $deg(D)$ or $deg(D')$ is zero, then the result is obvious. Suppose that $deg(D), deg(D') \geq 1$ and that the formulae hold whenever $deg(D) + deg(D') < k$. Choose $D, D' \in A_n$ such that $deg(D) + deg(D') = k$. It follows from (1) that it is enough to prove (2) and

(3) when D, D' are monomials. Suppose first that $D = \partial^\beta$ and $D' = x^\alpha$ with $|\alpha| + |\beta| = k$. If $\beta_i \neq 0$, then

$$[\partial^\beta, x^\alpha] = \partial_i[\partial^{\beta-e_i}, x^\alpha] + [\partial_i, x^\alpha]\partial^{\beta-e_i}.$$

By induction we have that

$$deg[\partial^{\beta-e_i}, x^\alpha] \leq |\alpha| + |\beta| - 3$$

and that $deg[\partial_i, x^\alpha] \leq |\alpha| - 1$. Hence we may use the induction hypothesis again to conclude that $deg(\partial_i[\partial^{\beta-e_i}, x^\alpha])$ and $deg([\partial_i, x^\alpha]\partial^{\beta-e_i})$ are $\leq |\alpha| + |\beta| - 2$. Therefore $deg[\partial^\beta, x^\alpha] \leq |\alpha| + |\beta| - 2$. But

$$\partial^\beta x^\alpha = [\partial^\beta, x^\alpha] + x^\alpha \partial^\beta.$$

Since $deg(x^\alpha \partial^\beta) = |\alpha| + |\beta|$ and $deg[\partial^\beta, x^\alpha] \leq |\alpha| + |\beta| - 2$, we conclude that

$$deg(\partial^\beta x^\alpha) = deg(x^\alpha \partial^\beta) = |\alpha| + |\beta|.$$

Now let $D = x^\sigma \partial^\beta$ and $D' = x^\alpha \partial^\eta$. If $|\alpha| = |\beta| = 0$, the result is obvious. Suppose that that is not the case. We have seen that $\partial^\beta x^\alpha = x^\alpha \partial^\beta + P$, where $P = [\partial^\beta, x^\alpha]$ has degree $\leq |\alpha| + |\beta| - 2$. Then

$$DD' = x^{\sigma+\alpha}\partial^{\beta+\eta} + x^\sigma P \partial^\eta.$$

By the induction hypothesis $deg(x^\sigma P \partial^\eta) \leq deg(D) + deg(D') - 2$. Hence

$$deg(DD') = deg(x^{\sigma+\alpha}\partial^{\beta+\eta}) = deg(D) + deg(D').$$

The calculations above also show that $D.D' = x^{\sigma+\alpha}\partial^{\beta+\eta} + Q_1$, where $deg(Q_1) \leq deg(D) + deg(D') - 2$. Similarly, we have that $D'D = x^{\sigma+\alpha}\partial^{\beta+\eta} + Q_2$, where $deg(Q_2) \leq deg(D) + deg(D') - 2$. Hence, $[D, D'] = Q_1 - Q_2$, and so $deg[D, D'] \leq deg(D) + deg(D') - 2$, which concludes the induction.

As in the case of polynomial rings over a field, Theorem 1.1(2) may be used to prove the following result.

1.2 COROLLARY. *The algebra A_n is a domain.*

2. IDEAL STRUCTURE.

If you are used to commutative rings, you will find A_n very peculiar. Commutative rings have lots of two-sided ideals; not so A_n. A ring whose only proper ideal is zero is called *simple*. A commutative simple ring must be a field, but this is not true of noncommutative rings. The Weyl algebra is a simple ring, but it is very far from being even a division ring.

2.1 THEOREM. *The algebra A_n is simple.*

PROOF: Let I be a non-zero two-sided ideal of A_n. Choose an element $D \neq 0$ of smallest degree in I. If D has degree 0, it is a constant, and $I = A_n$. Assume that D has degree $k > 0$ and let us aim at a contradiction.

Suppose that (α, β) is a multi-index of length k. If $x^\alpha \partial^\beta$ is a summand of D with non-zero coefficient and $\beta_i \neq 0$, then $[x_i, D]$ is non-zero and has degree $k-1$, by Theorem 1.1. Since I is a two-sided ideal of A_n, we have that $[x_i, D] \in I$. But this contradicts the minimality of D. Thus $\beta = (0, \ldots, 0)$. Since $k > 0$, we must have that $\alpha_i \neq 0$, for some $i = 1, 2, \ldots, n$. Hence $[\partial_i, D]$ is a non-zero element of I of degree $k - 1$, and once again we have a contradiction.

The kernel of an endomorphism of A_n is a two-sided ideal; thus we have the following corollary.

2.2 COROLLARY. *Every endomorphism of A_n is injective.*

Although A_n does not have any non-trivial two-sided ideals, it is not a division ring. In fact, the only elements of A_n that have an inverse are the constants. This is easy to see, as for polynomial rings, using the degree of an operator. If $D \in A_n$ has an inverse, then there exists $D' \in A_n$ such that $DD' = 1$. Calculating degrees, we obtain $deg(D) + deg(D') = 0$. Since the degree of a non-zero operator is always non-negative, we must have $deg(D) = 0$. Hence D is constant. Thus every non-constant operator generates a non-trivial left ideal of A_n.

The Weyl algebra is not a left principal ideal ring either. For example, the left ideal generated by ∂_1, ∂_2 in A_2 is not principal; see Exercise 4.1. However,

2. *Ideal structure* 17

every left ideal of A_n is generated by two elements. This is a very important result, due to J.T. Stafford. Unfortunately, its proof is very technical and beyond the scope of this book. It may be found in the original paper of Stafford [Stafford **78**] and also in [Björk **79**, Ch. 1]. In fact, a similar result holds for left ideals over any ring of differential operators over a smooth variety, see [Coutinho and Holland **88**].

3. POSITIVE CHARACTERISTIC.

At the beginning of chapter 1 we made the hypothesis that the base field K over which the Weyl algebra is defined always has characteristic zero. However, it is not immediately clear why one should make such a restriction. After all, the definition of the Weyl algebra makes perfect sense without any restriction on the characteristic of the field K.

The problem lies deeper. First of all, the Weyl algebra in positive characteristic suffers from a split personality problem. This happens because the two ways by which we defined A_n in Ch. 1 give two non-isomorphic rings. Let us see what happens for the field $\mathbb{Z}_p$ where p is a prime, in the one-variable case.

Consider first the algebra of operators R_1 generated by $\mathbb{Z}_p[x]$ and its derivation ∂. Let us calculate $\partial^p(x^k)$. If $k < p$, then $\partial^p(x^k) = 0$. If $k \geq p$, then

$$\partial^p(x^k) = k \ldots (k - p + 1)x^{k-p} = 0$$

since the coefficient is divisible by p. Thus $\partial^p = 0$ as an operator on $\mathbb{Z}_p[x]$. We conclude that the ring of operators R_1 has nilpotent elements. In particular it is not a domain.

Now consider the ring R_2 generated over $\mathbb{Z}_p$ by z_1, z_2 subject to the relation $[z_2, z_1] = 1$. This is a domain. However, it is not like the Weyl algebra in characteristic zero in another respect: it is not a simple ring. For example, if $f \in \mathbb{Z}_p[z_1]$ then $[z_2, f] = \partial f / \partial z_1$. In particular,

$$[z_2, z_1^p] = pz_1^{p-1} = 0$$

over $\mathbb{Z}_p$. It follows that z_1^p commutes with every element of R_2. Hence the left ideal generated by z_1^p in R_2 is a two-sided ideal. In particular R_2 is not a simple ring.

These superficial comments are enough to show that the positive characteristic case is very different from the characteristic zero case which is the
subject of this book. The Weyl algebra in positive characteristic is studied
in great detail in [Smith 86].

4. Exercises.

4.1 Let $n \geq 2$. Show that the left ideal L of A_n generated by $\partial_1, \ldots, \partial_n$ is
not principal.

Hint: Show that a generator would have degree 1, and obtain a contradiction
from that.

4.2 Show that the left ideal of A_3 generated by $\partial_1, \partial_2, \partial_3$ may also be generated
by two elements.

Hint: Choose $D_1 = \partial_1$ and $D_2 = \partial_2 + x_1 \partial_3$ to be the generators, and calculate
$[D_1, D_2]$.

4.3 Let $K(X)$ be the rational function field in n variables. Let $B_n(K)$ be the
K-subalgebra of $End_K(K(X))$ generated by the elements of $K(X)$ and the
derivations $\partial_1, \ldots, \partial_n$. An element $d \in B_n(K)$ may be written in the form
$d = \sum_\alpha q_\alpha \partial^\alpha$, where $q_\alpha \in K(X)$. The *order* of d, denoted by $ord(d)$, is the
largest $|\alpha|$ for which $q_\alpha \neq 0$. Show that if $d' \in B_n(K)$ then:

   (1) $ord(d + d') \leq \max\{ord(d), ord(d')\}$.
   (2) $ord(dd') = ord(d) + ord(d')$.

4.4 Use the order defined above to show that:

   (1) $B_n(K)$ is a domain.
   (2) $B_n(K)$ is a simple ring.

4.5 Show that $B_1(K)$ admits a left division algorithm. That is, if $a, b \in$
$B_1(K)$, then there exist $q, r \in B_1(K)$ such that $a = qb + r$ and $ord(r) <$
$ord(b)$. Use this to prove that every left ideal of $B_1(K)$ is principal.

4.6 If $n \geq 2$, is every left ideal of $B_n(K)$ principal ?

4.7 Let $A_1(\mathbb{Z})$ be the $\mathbb{Z}$-subalgebra of $A_1(\mathbb{Q})$ generated by x and ∂. Show that $A_1(\mathbb{Z})$ is not a simple ring.

4.8 Let $d \in A_1(K)$, and write $d = \sum_0^m g_i(x)\partial^i$. Call m the *order* of d and $L(d) = g_m(x) \neq 0$ its *leading term*. If J is a left ideal of $A_1(K)$, show that:

(1) The set $L_n(J) = \{L(d) : d \in J \text{ and } d \text{ has order } n\} \cup \{0\}$ is an ideal of $K[x]$.

(2) If $m \leq n$, then $L_m(J) \subseteq L_n(J)$.

(3) J is principal if and only if $L_n(J) = L_{n+1}(J)$ for all $n > 0$ such that $L_n(J) \neq 0$.

4.9 Show that the left ideal $A_1\partial^2 + A_1(x\partial - 1)$ is not a cyclic ideal of $A_1(K)$.

4.10 Let $\phi : A_1^2 \to A_1$ be the map defined by $\phi(a, b) = a\partial + bx$. Show that ϕ is surjective and that its kernel is isomorphic to the left ideal $A_1\partial^2 + A_1(x\partial - 1)$. Conclude that this is a projective left ideal of $A_1(K)$.

4.11 Let R_2 be the ring defined in §3. Show that:

(1) R_2 is a domain.

(2) The centre of R_2 is $\mathbb{Z}_p[z_1^p, z_2^p]$.

(3) R_2 is finitely generated as a module over its centre.

CHAPTER 3

RINGS OF DIFFERENTIAL OPERATORS

In this chapter we show that the Weyl algebras are members of the family of rings of differential operators. These rings come up in many areas of mathematics: representation theory of Lie algebras, singularity theory and differential equations are some of them.

1. DEFINITIONS.

Let R be a commutative K-algebra. The ring of differential operators of R is defined, inductively, as a subring of $End_K(R)$. As in the case of the Weyl algebra, we will identify an element $a \in R$ with the operator of $End_K(R)$ defined by the rule $r \longmapsto ar$, for every $r \in R$.

We now define, inductively, the order of an operator. An operator $P \in End_K(R)$ has *order zero* if $[a, P] = 0$, for every $a \in R$. Suppose we have defined operators of order $< n$. An operator $P \in End_K(R)$ has *order n* if it does not have order less than n and $[a, P]$ has order less than n for every $a \in R$. Let $D^n(R)$ denote the set of all operators of $End_K(R)$ of order $\leq n$. It is easy to check, from the definitions, that $D^n(R)$ is a K-vector space.

We may characterize the operators of order ≤ 1 in terms of well-known concepts. A *derivation* of the K-algebra R is a linear operator D of R which satisfies Leibniz's rule: $D(ab) = aD(b) + bD(a)$ for every $a, b \in R$. Let $Der_K(R)$ denote the K-vector space of all derivations of R. Of course $Der_K(R) \subseteq End_K(R)$. If $D \in Der_K(R)$ and $a \in R$, we define a new derivation aD by $(aD)(b) = aD(b)$ for every $b \in R$. The vector space $Der_K(R)$ is a left R-module for this action. Derivations are very important in commutative algebra and algebraic geometry as shown in [Matsumura **86**] and [Hartshorne **77**].

1.1 LEMMA. *The operators of order ≤ 1 correspond to the elements of $Der_K(R) + R$. The elements of order zero are the elements of R.*

PROOF: Let $Q \in D^1(R)$ and put $P = Q - Q(1)$. Note that $P(1) = 0$ and that P has order ≤ 1. Hence $[P, a]$ has order zero for every $a \in R$. Thus for every $b \in R$, we have that $[[P, a], b] = 0$. Writing the commutators explicitly, one obtains the equality

$$(Pa)b - (aP)b - b(Pa) + b(aP) = 0.$$

Applying this operator to $1 \in R$, we end up with $P(ab) = aP(b) + bP(a) - baP(1)$. Since $P(1) = 0$, it follows that P is a derivation of R. But $Q = P + Q(1) \in Der_K(R) + R$, as required. An easy calculation shows that if Q has order zero then $Q \in R$.

The *ring of differential operators* $D(R)$ of the K-algebra R is the set of all operators of $End_K(R)$ of finite order, with the operations of sum and composition of operators. In other words, $D(R)$ is the union of the K-vector spaces $D^n(R)$ for $n = 0, 1, 2 \ldots$. For this definition to make sense, we must show that the sum and product of two operators of finite order has finite order. This is clear for the sum, but requires a proof for the multiplication.

1.2 PROPOSITION. *Let* $P \in D^n(R)$ *and* $Q \in D^m(R)$, *then* $P \cdot Q \in D^{n+m}(R)$.

PROOF: The proof is by induction on $m + n$. If $m + n = 0$ the result is obvious. Suppose the result true whenever $m + n < k$. If $m + n = k$ and $a \in R$, we have that

$$[PQ, a] = P[Q, a] + [P, a]Q.$$

The definition of order implies that $[Q, a] \in D^{m-1}(R)$ and $[P, a] \in D^{n-1}(R)$. Thus, by the induction hypothesis $P[Q, a], [P, a]Q \in D^{n+m-1}$. Hence $[PQ, a]$ belongs to D^{n+m-1}, as required.

We end this section with the explicit calculation of $Der_K(K[x_1, \ldots, x_n])$, which will be needed in the next section.

1.3 PROPOSITION. *Every derivation of* $K[X] = K[x_1, \ldots, x_n]$ *is of the form* $\sum_1^n f_i \partial_i$, *for some* $f_1, \ldots, f_n \in K[X]$.

PROOF: Let $D \in Der_K(K[X])$. Then $D(x_i^k) = kx_i^{k-1}D(x_i)$, for $i = 1, \ldots, n$. Hence

$$(D - \sum_1^n D(x_i)\partial_i)(x_1^{s_1} \ldots x_n^{s_n}) = 0.$$

Since these monomials form a basis for $K[X]$ we have that $D = \sum_1^n D(x_i)\partial_i$. Thus $D = \sum_1^n f_i \partial_i$ with $f_i = D(x_i)$.

2. The Weyl algebra.

Our aim now is to show that the Weyl algebra is the ring of differential operators of the algebra of polynomials. For the proof of this result we need two lemmas.

2.1 Lemma. *Let $P \in D(K[X])$. If $[P, x_i] = 0$ for every $i = 1, \ldots, n$, then $P \in K[X]$.*

PROOF: We will show that under this hypothesis, $[P, f] = 0$ for every $f \in K[X]$. Then the result follows from Lemma 1.1. Since the commutator is additive, it is enough to prove that $[P, f] = 0$ when f is a monomial in $K[X]$. Let $f = x^\alpha$, for some $\alpha \in \mathbb{N}^n$ and assume that $\alpha_i \neq 0$. Then

$$[P, x^\alpha] = [P, x_i]x^{\alpha - e_i} + x_i[P, x^{\alpha - e_i}].$$

By induction on the degree of the monomials, $[P, x_i] = [P, x^{\alpha - e_i}] = 0$. Thus $[P, x^\alpha] = 0$, as required.

The next lemma is formally equivalent to the fact that every polynomial vector field $F = (F_1, \ldots, F_n)$ in $\mathbb{R}^n$ which satisfies $\partial F_j / \partial x_i = \partial F_i / \partial x_j$ for all $1 \leq i, j \leq n$, has a potential. Define C_r to be the set of operators in A_n which can be written in the form $\sum_\alpha f_\alpha \partial^\alpha$ with $|\alpha| \leq r$. A simple calculation shows that

$$C_r = C_{r+1} \cap D^r(K[X]).$$

By Proposition 1.3, we have that $C_1 = Der_K(K[X]) + K[X]$ and that $C_0 = K[X]$. We will use the convention that if $k < n$ then $\mathbb{N}^k$ is embedded in $\mathbb{N}^n$ as the set of n-tuples whose last $n - k$ components are zero.

2.2 Lemma. *Let $P_1, \ldots, P_n \in C_{r-1}$ and assume that $[P_i, x_j] = [P_j, x_i]$ whenever $1 \leq i, j \leq n$. Then there exists $Q \in C_r$ such that $P_i = [Q, x_i]$, for $i = 1, \ldots, n$.*

PROOF: Suppose, by induction, that we have determined $Q' \in C_r$ such that $[Q', x_i] = P_i$ for $k + 1 \leq i \leq n$; thus $[[Q', x_i], x_k] = [P_k, x_i]$. Write $G = [Q', x_k] - P_k$. Then $[G, x_i] = 0$, for $k + 1 \leq i \leq n$.

It follows from the identity

$$[\partial^\beta, x_n] = \beta_n \partial^{\beta-e_n}$$

that if $[\partial^\beta, x_n] = 0$ then $\beta_n = 0$. Thus $[G, x_n] = 0$ implies that G can be written as a linear combination of monomials of the form $x^\alpha \partial^\beta$ with $\beta \in \mathbb{N}^{n-1}$. Since $[G, x_i] = 0$ for $k+1 \le i \le n$, we may apply this result several times and conclude that

$$G = \sum_{\alpha \in \mathbb{N}^k} f_\alpha \partial^\alpha$$

where $f_\alpha \in K[X]$. Now write

$$Q'' = \sum_{\alpha \in \mathbb{N}^k} (\alpha_k + 1)^{-1} f_\alpha \partial^{\alpha+e_k}.$$

However, $Q' \in C_r \subseteq D^r(K[X])$ implies that

$$[Q', x_k] \in C_r \cap D^{r-1}(K[X]) = C_{r-1}.$$

Since P_k also belongs to C_{r-1}, then so does G. Hence $Q'' \in C_r$. On the other hand, $[Q'', x_i] = 0$ for $k+1 \le i \le n$ by construction. Thus $[Q' - Q'', x_i] = P_i$. But $[Q'', x_k] = G$, and so

$$[Q' - Q'', x_k] = [Q', x_k] - G = P_k$$

Hence, $[Q' - Q'', x_i] = P_i$, for $k \le i \le n$; and the induction is complete.

2.3 THEOREM. *The ring of differential operators of $K[X]$ is $A_n(K)$. Besides this, $D^k(K[X]) = C_k$.*

PROOF: It is enough to prove that $D^k(K[X]) \subseteq C_k$. Let $P \in D(K[X])$. If $P \in D^1(K[X])$ then by Lemma 1.1, $P \in Der_K(K[X]) + K[X]$. Thus $P \in C_1$ by Proposition 1.3. Suppose, by induction, that $D^k(K[X]) = C_k$ for $k \le m-1$. Let $P \in D^m(K[X])$. Write $P_i = [P, x_i]$. Since P_i has order $k \le m-1$, it follows that $P_i \in C_{m-1}$. But, for all $1 \le i, j \le n$,

$$[P_i, x_j] = [[P, x_i], x_j] = [[P, x_j], x_i] = [P_j, x_i].$$

Thus by Lemma 2.2 there exists $Q \in C_m$ such that $[Q, x_i] = P_i$, $1 \le i \le n$. Hence $[Q - P, x_i] = 0$ in $D(K[X])$. Since this holds whenever $1 \le i \le n$, we conclude by Lemma 2.1 that $Q - P \in K[X] = C_0$. Therefore $P \in C_m$. Hence $D^m(K[X]) \subseteq C^m$, as we wanted to prove.

The ring of differential operators $D(R)$ of a commutative domain R is not always generated by R and $Der_K(R)$; see Exercise 3.8. However this is true if the ring R is regular; for example, if it is the coordinate ring of a non-singular irreducible affine variety. For a proof of this see [McConnell and Robson **87**, Ch.15]. The ring of Exercise 3.8 is the coordinate ring of a *singular* curve, the cusp.

3. EXERCISES.

3.1 Let J be a right ideal of a ring R. The idealizer of J in R is the set $\mathbb{I}(J) = \{a \in R : aJ \in J\}$. Show that $\mathbb{I}(J)$ is the largest subring of R that contains J as a two-sided ideal.

3.2 Let J be an ideal of $S = K[x_1, \ldots, x_n]$. Let $\theta \in End_K S$.

(1) Show that the formula $\bar{\theta}(\bar{f}) = \overline{\theta(f)}$ defines a K-endomorphism of S/J if and only if $\theta(J) \subseteq J$. In that case, show that $\bar{\theta} = 0$ if and only if $\theta(S) \subseteq J$.

(2) Show that if $\theta \in A_n \subseteq End_K S$ and $\theta(J) \subseteq J$, then $\bar{\theta} \in D(S/J)$, the ring of differential operators of the quotient S/J. Show also that the order of $\bar{\theta}$ cannot exceed the order of θ.

(3) Show that if θ is a derivation of S then $\bar{\theta}$ is a derivation of S/J.

3.3 Let J be an ideal of $S = K[x_1, \ldots, x_n]$. Let $P \in A_n$. Show that:

(1) If $P(S) \subseteq J$ then $P \in JA_n$.

(2) If $P(J) \subseteq J$ then $P \in \mathbb{I}(JA_n)$.

3.4 Let J be an ideal of $S = K[x_1, \ldots, x_n]$. Let $D(S/J)$ be the ring of differential operators of the quotient ring S/J. Show that there is an injective ring homomorphism from $\mathbb{I}(JA_n)/JA_n$ into $D(S/J)$.

Hint: There is a homomorphism from $\mathbb{I}(JA_n)$ to $D(S/J)$ which maps θ to $\bar{\theta}$, in the notation of Exercise 3.2. Now apply Exercise 3.3.

3.5 Let J be an ideal of $S = K[x_1, \ldots, x_n]$. Let $\bar{f}$ be the image of $f \in S$ in the quotient ring S/J. Suppose that D is a derivation of S/J and choose $g_i \in S$ such that $\bar{g_i} = D(\bar{x_i})$, for $1 \le i \le n$.

   (1) Show that if $B = \sum_1^n g_i \partial_i$, then $\overline{B} = D$, in the notation of Exercise 3.2.

   (2) Let $Der_J(S)$ be the set of derivations $D \in Der(S)$ such that $D(J) \subseteq J$. Conclude, using Exercise 3.2, that there is an isomorphism of vector spaces between $Der_J(S)/JDer_J(S)$ and $Der(S/J)$.

3.6 Let $R = K[t^2, t^3]$.

   (1) Show that R is isomorphic to $K[x, y]/J$, where J is the ideal of $K[x, y]$ generated by $y^2 - x^3$.

   (2) Let $D_1 = 2y\partial_x + 3x^2\partial_y$ and $D_2 = 3y\partial_y - 2x\partial_x$. Use the previous exercise to show that the set of derivatives of $K[x, y]/J$ is generated by D_1 and D_2 as a module over $K[x, y]/J$.

   (3) Conclude that $Der_K(R)$ is an R-module generated by $t\partial_t$ and $t^2\partial_t$, where $\partial_t = d/dt$.

3.7 Let R be a commutative domain with field of fractions Q. Show that the set $\{P \in D(Q) : P(R) \subseteq R\}$ is a subring of $D(R)$.

3.8 Let $R = K[t^2, t^3]$. Let $\partial = d/dt$ and let B_1 be the K-algebra generated by $K(t)$ and ∂; see Exercise 2.4.3.

   (1) Using Exercise 3.7, show that the following elements of B_1 belong to $D(R)$: $\partial^2 - 2t^{-1}\partial$, $t\partial^2 - \partial$ and $\partial^3 - 3t^{-1}\partial^2 + 3t^{-2}\partial$.

   (2) Show that these elements do not belong to the subring of B_1 generated by R and its derivations.

   (3) Conclude that $D(R)$ is *not* generated by R and $Der_K(R)$.

CHAPTER 4
JACOBIAN CONJECTURE

The Jacobian conjecture was proposed by O.H. Keller in 1939. It asks whether a polynomial endomorphism of $\mathbb{C}^n$ whose Jacobian is constant must be invertible. Despite its simple and reasonable statement, the conjecture has not been proved even in the two dimensional case. In this chapter we show that this conjecture would follow if one could prove that every endomorphism of the Weyl algebra is an automorphism. The chapter opens with a discussion of polynomial maps, which will play a central rôle in the second part of the book. We shall return to the Jacobian conjecture in Ch. 19.

1. POLYNOMIAL MAPS.

Let $F : K^n \to K^m$ be a map and p a point of K^n. We say that F is *polynomial* if there exist $F_1, \ldots, F_m \in K[x_1, \ldots, x_n]$ such that $F(p) = (F_1(p), \ldots, F_m(p))$. A polynomial map is called an *isomorphism* or a *polynomial isomorphism* if it has an inverse which is also a polynomial map. It is not always the case that a bijective polynomial map has an inverse which is also polynomial. For an example where this does not occur see Exercise 5.1. However, if $K = \mathbb{C}$, every invertible polynomial map has a polynomial inverse. This is proved in [Bass, Connell and Wright **82**; Theorem 2.1].

For the rest of the section we shall write X, Y for the spaces K^n and K^m; and $K[X]$, $K[Y]$ for the polynomial rings $K[x_1, \ldots, x_n]$ and $K[y_1, \ldots, y_m]$.

A polynomial $g \in K[Y]$ may be identified with the function of Y into K which maps $p \in Y$ to $g(p)$. Clearly if $g = 0$ as a polynomial, then it induces the zero function on Y. Since K is a field of characteristic zero, the converse is also true: a polynomial which induces the zero function on Y is identically zero. For a proof see Exercise 5.2. This identification is the key to the construction that follows.

Suppose that $F : X \to Y$ is a polynomial map. We may define a map,

$$F^\sharp : K[Y] \to K[X],$$

by the formula $F^\sharp(g) = g \cdot F$, where $g \in K[Y]$. Note that the arrow gets reversed as we go from F to $F^\sharp$. The map $F^\sharp$ is called the *comorphism* of F. A routine calculation shows that $F^\sharp$ is a homomorphism of polynomial rings. Let us calculate an example. Suppose that $n < m$ and that $F : X \to Y$ is the map

$$F(x_1, \dots, x_n) = (x_1, \dots x_n, 0, \dots, 0).$$

Then the algebra homomorphism $F^\sharp : K[Y] \to K[X]$ maps a polynomial $g(y_1, \dots, y_m)$ to $g(x_1, \dots, x_n, 0, \dots, 0)$.

We may turn the construction of the above paragraph inside out. Suppose that a ring homomorphism $\phi : K[Y] \to K[X]$ is given. Then we may use it to construct a polynomial map from X to Y. Let y_i be an indeterminate in $K[Y]$. Then $\phi(y_i)$ is a polynomial of $K[X]$. Now let $\phi_\sharp : X \to Y$ be the map whose coordinate functions are $\phi(y_1), \dots, \phi(y_m)$. These two constructions are each other's inverse, as the next results show.

1.1 THEOREM. *Let $F : X \to Y$ be a polynomial map, then $(F^\sharp)_\sharp = F$. Furthermore, if $Z = K^r$ and $G : Y \to Z$ is another polynomial map, then $G \cdot F : X \to Z$ is a polynomial map and $(G \cdot F)^\sharp = F^\sharp \cdot G^\sharp$.*

PROOF: Let y_i be an indeterminate in $K[Y]$. As a function $Y \to K$ we have that y_i maps a point of Y onto its i-th coordinate. Hence,

$$F^\sharp(y_i) = y_i \cdot F = F_i.$$

Thus, the coordinate functions of $(F^\sharp)_\sharp$ are $F_1, \dots, F_m$; which are the coordinate functions of F. Hence $(F^\sharp)_\sharp = F$.

It is clear that $G \cdot F$ is a polynomial function. Let $g \in K[Z] = K[z_1, \dots, z_r]$. Then

$$(G \cdot F)^\sharp(g) = g \cdot (G \cdot F).$$

Since the composition of maps is associative,

$$(G \cdot F)^\sharp(g) = (g \cdot G) \cdot F = F^\sharp(G^\sharp(g)),$$

as required.

The converse of Theorem 1.1 is also true. We state it and leave the proof to the reader.

1.2 THEOREM. *If $\phi : K[Y] \to K[X]$ is a homomorphism of polynomial rings,*
then $(\phi_\sharp)^\sharp = \phi$. Furthermore, if $\psi : K[Z] \to K[Y]$ is another homomorphism,
then

$$(\phi \cdot \psi)_\sharp = \psi_\sharp \cdot \phi_\sharp.$$

The following result is an immediate consequence of Theorems 1.1 and
1.2.

1.3 COROLLARY. *A polynomial map $F : X \to Y$ is an isomorphism if and*
only if $F^\sharp$ is an isomorphism.

The Jacobian conjecture, discussed in the next section, proposes a simple
criterion to determine whether a polynomial map is an isomorphism.

2. JACOBIAN CONJECTURE

Let $F : K^n \to K^n$ be a polynomial map. Denote by $J(F)$ its jacobian
matrix. This is the matrix whose ij entry is $\partial F_i / \partial x_j$. If F is an isomorphism,
then it follows from the chain rule that $J(F)$ is an invertible matrix at every
point of K^n. In particular the determinant $\Delta F = det J(F)$ is an invertible
polynomial, hence a constant.

Now suppose that $K = \mathbb{R}$ or $\mathbb{C}$. If $p \in K^n$ and $\Delta F(p) \neq 0$, then by the
inverse function theorem there exists a neighbourhood U of p such that F
restricted to U is invertible. This leads to the following question. If ΔF
is non-zero everywhere on K^n, is there a function $G : K^n \to K^n$ which is
the inverse of F ? Two comments are in order. First, although F always
has an inverse in the neighbourhood of every point (by the inverse function
theorem) it is not clear whether these inverses can be 'glued' together to
produce an inverse in the whole of K^n. Secondly, even if the inverse exists,
it may not be a polynomial map, as shown in Exercise 5.1. The Jacobian
conjecture, first stated in [Keller **39**], is a refinement of the above question.

2.1 JACOBIAN CONJECTURE. *Let $F : K^n \to K^n$ be a polynomial map. If*
$\Delta F = 1$ on K^n then F has a polynomial inverse on the whole of K^n.

Let us see what happens if $n = 1$. In this case, we have a map $F : K \to K$
which is determined by a polynomial $F(x)$ in one variable. The jacobian
matrix is the derivative dF/dx, and we are assuming that this is a constant.

Hence F is a linear map, and consequently it has an inverse. This proves that the Jacobian conjecture holds for $n = 1$. The fact that an invertible polynomial is linear is restricted to the one-dimensional case. For $n \geq 2$, an invertible polynomial map may have coordinate functions of any degree whatsoever; as shown by the examples of Exercise 5.3.

Despite many attempts to settle it, the Jacobian conjecture remains open for every $n \geq 2$. A number of results related to the Jacobian conjecture have accumulated over the years; for a very nice survey of some of these results see [Bass, Connell and Wright 82]. For example, it is known that the conjecture is false over fields of positive characteristic, see Exercise 5.4. On the positive side, if all the coordinate functions of F have degree ≤ 2, then the conjecture is true; see Exercise 5.5.

We now rephrase the Jacobian conjecture using comorphisms. Let $X = K^n$ and $K[X] = K[x_1, \ldots, x_n]$. The Jacobian Conjecture states that if $\Delta F = 1$ on X then $F^{\sharp}$ is an invertible homomorphism of rings. Actually it is not hard to see that if ΔF is constant, then $F^{\sharp}$ is necessarily injective. We prove this in a little more generality.

2.2 LEMMA. *Let $F : X \to X$ be a polynomial map and suppose that $\Delta F \neq 0$ everywhere in X. Then $F^{\sharp}$ is injective.*

PROOF: Suppose that $F^{\sharp}$ is not injective, and choose the non-constant polynomial $g \in K[X]$ of smallest degree such that $F^{\sharp}(g) = 0$. Then $g(F) = 0$. Let $g_i = \partial g / \partial x_i$ and

$$\mathbf{v} = (g_1(F_1, \ldots, F_n), \ldots, g_n(F_1, \ldots, F_n)).$$

Hence, by the chain rule,

$$\mathbf{v}(p) \cdot JF(p) = 0$$

for every $p \in X$. Since

$$\Delta F(p) = det JF(p) \neq 0,$$

we conclude that $\mathbf{v}(p) = 0$ for every $p \in X$. Thus $g_i(F_1, \ldots, F_n) = 0$ for $1 \leq i \leq n$. Since g is not constant, at least one of the g_i must be non-zero. But g_i has degree smaller than g, a contradiction.

Denote by $K[F_1, \ldots, F_n]$ the subalgebra of $K[X]$ generated by the coordinate functions of F. This is the image of the homomorphism $F^\sharp$. If $\Delta F = 1$ then $F^\sharp$ is injective by Lemma 2.2. Now assume that $K[F_1, \ldots, F_n] = K[X]$; then $F^\sharp$ is also surjective. Hence by Corollary 1.3, F itself is bijective. Thus the Jacobian conjecture may be rephrased as follows.

2.3 JACOBIAN CONJECTURE. *Let* $F : K^n \to K^n$ *be a polynomial map and assume that* $\Delta F = 1$ *in* K^n. *Then* $K[F_1, \ldots, F_n] = K[x_1, \ldots, x_n]$.

In §4 we show that if every endomorphism of the Weyl algebra is an automorphism then the Jacobian conjecture holds. This follows an idea of L. Vaserstein and V. Katz; see [Bass, Connell and Wright **82**, p.297]. The next section is a digression on results about derivations that will be required in §4.

3. DERIVATIONS

In this section we study some properties of derivations that will be required in the next section. Let D be a derivation of a K-algebra S. It follows from Leibniz's rule that the kernel of D is a subring of S, it is called the *ring of constants* of D. The derivation D is *locally nilpotent* if given $a \in S$, there exists $k \in \mathbb{N}$ such that $D^k(a) = 0$. Note that the derivations $\partial_1, \ldots, \partial_n$ are locally nilpotent in $K[x_1, \ldots x_n]$, whilst $x_1 \partial_1$ is not.

Let S be a ring and D a locally nilpotent derivation. Define a map $\phi : S \longrightarrow S[x]$ by the rule

$$\phi(a) = \sum_0^\infty \frac{D^n(a)}{n!} x^n$$

for every $a \in S$. Note that $\phi(a)$ belongs to $S[x]$ because D is locally nilpotent. It is easy to check that ϕ is a ring homomorphism which satisfies

$$\phi \cdot D = \frac{d}{dx} \cdot \phi.$$

We want to prove the following proposition from [Wright **81**].

PROPOSITION 3.1. *Let S be a K-algebra and $D_1, \ldots, D_n$ be commuting locally nilpotent derivations of S. Suppose that there exist $t_1, \ldots, t_n \in S$ such that $D_i(t_j) = \delta_{ij}$. Then*

(1) $S = R[t_1, \ldots, t_n]$, *where R is the ring of constants with respect to $D_1, \ldots, D_n$,*

(2) $t_1, \ldots, t_n$ *are algebraically independent over R,*

(3) $D_i = \partial/\partial t_i$ *for $i = 1, \ldots, n$.*

The proposition is proved by induction. It is better to isolate the case $n = 1$ in a lemma.

LEMMA 3.2. *Let S be a K-algebra and D a locally nilpotent derivation of S. Suppose that for some $t \in S$ one has $D(t) = 1$. Then*

(1) $S = R[t]$, *where R is the ring of constants of S,*

(2) t *is algebraically independent over R,*

(3) $D = d/dt$.

PROOF: Put $\bar{S} = S/St$. Let $\rho : S \longrightarrow \bar{S}[x]$ be the composition of ϕ defined above and the projection $S[x] \longrightarrow \bar{S}[x]$. We want to show that ρ is an isomorphism. Note that $\rho(t) = x$.

To prove that ρ is surjective it is enough to prove that its image contains $\bar{S}$. Let $a \in S$. Denote by $\bar{a}$ its image in $\bar{S}$. Since D is locally nilpotent, there exists $n \in \mathbb{N}$ such that $D^k(a) = 0$ for $k > n$. Thus,

$$\rho(a) = \sum_0^n \frac{\overline{D^i(a)}}{i!} x^i.$$

If $n = 0$, then $\rho(a) = \bar{a}$. If $n > 0$ put $a_0 = a$ and define $a_{j+1} = a_j - D^{n-j}(a_j)t^{n-j}/(n-j)!$, for $j = 1, \ldots, n$. It is easy to show, by induction on j, that $D^k(a_j) = 0$ for $k > n - j$ and that

$$\rho(a_j) = \sum_0^{n-j} \frac{\overline{D^i(a_j)}}{i!} x^i.$$

Thus $\rho(a_n) = \bar{a}_n$. However, since $\bar{t} = 0$, we have that $\rho(a_n) = \bar{a}$. Thus ρ is surjective.

Let us prove that ρ is injective. If not, then there exists a non-zero $a \in S$ such that $\rho(a) = 0$. Thus $D^k(a) \in tS$, for every $k \in \mathbb{N}$. Hence $a = a_1 \cdot t$, for some $a_1 \in S$. Since $\rho(t) = x$, we have that $\rho(a_1) = 0$. Thus $a_1 \in tS$ and $a = a_2 \cdot t^2$, for some $a_2 \in S$. Continuing this way we conclude that t^n divides a for all $n \geq 0$. But this is impossible, unless $a = 0$. Indeed, ϕ maps t to $t + x$. Thus if t^n divides a, we also have that $\phi(t^n) = (t + x)^n$ divides $\phi(a)$ in the polynomial ring $S[x]$. Hence, if $a \neq 0$ we have that $deg(\phi(a)) \geq n$ for every $n > 0$, which is clearly impossible. Thus $a = 0$, as required.

We conclude that the homomorphism $\rho : S \longrightarrow \overline{S}[x]$ is an isomorphism. Since $\rho \cdot D = d/dx \cdot \rho$, we have that $R = \rho^{-1}(\overline{S})$. The result now follows if we recall that $\rho(t) = x$.

PROOF OF PROPOSITION 3.1: We proceed by induction on the number n of derivations. By Lemma 3.2, $S = R_1[t_1]$, where R_1 is the ring of constants of D_1. But t_1 is algebraically independent over R_1 and $D_1 = d/dt_1$. Since D_1 commutes with D_i for $i > 1$, we have that $D_i(R_1) \subseteq R_1$. Thus, by the induction hypothesis, $R_1 = R[t_2, \ldots t_n]$, and the proposition follows.

4. AUTOMORPHISMS

We now return to the setup of the Jacobian conjecture. Let $X = K^n$. The rational function field of $K[X]$ will be denoted by $K(X)$. Let $F : X \to X$ be a polynomial map with coordinate functions $F_1, \ldots, F_n$. Assume that

$$\Delta = \Delta F \neq 0$$

everywhere on X; a condition that is actually weaker than the one required by the Jacobian conjecture. Define a map $D_i : K(X) \to K(X)$ by

$$D_i(g) = \Delta^{-1} det J(F_1, \ldots, F_{i-1}, g, F_{i+1}, \ldots, F_n).$$

It is easy to check that D_i is a K-linear map that satisfies Leibniz's rule. Hence D_i is a derivation of $K(X)$. Now let $K[X, \Delta^{-1}]$ be the K-subalgebra of $K(X)$ of all rational functions whose denominator is a power of Δ. Then D_i restricts to a derivation of $K[X, \Delta^{-1}]$, since

$$D_i(\Delta^{-1}) = -\Delta^{-2} D_i(\Delta).$$

4.1 LEMMA. *As derivations of $K[X, \Delta^{-1}]$ the D_i satisfy:*

(1) $D_i(F_j) = \delta_{ij}$.

(2) *The D_i commute pairwise.*

PROOF: We prove (2); since (1) follows easily from properties of the determinant. Note first that $\Delta(0) \neq 0$. Thus Δ is invertible as a power series and $K[X, \Delta^{-1}] \subseteq K[[X]]$. On the other hand, $\Delta \cdot D_i$ is a derivation of $K[x_1, \ldots, x_n]$ which can be extended to a derivation on the power series ring $K[[X]] = K[[x_1, \ldots, x_n]]$; see Exercise 5.9. Since Δ is invertible as a power series, then D_i can also be extended to a derivation of $K[[X]]$.

Put $B = [D_i, D_j]$. We want to show that $B = 0$ on $K[X, \Delta^{-1}]$. It is enough to show that $B = 0$ on the power series ring $K[[X]]$. Since the commutator of two derivations is a derivation (see Exercise 5.8), we have that B is a derivation of $K[[X]]$. Moreover $B(F_k) = 0$, for $1 \leq k \leq n$; and so B is zero in the subalgebra $K[F_1, \ldots, F_n]$. But $F_1, \ldots, F_n$ are algebraically independent, by Lemma 2.2. Hence we may consider B as a derivation on the power series ring $K[[F_1, \ldots, F_n]]$. By (1), B is zero on $K[[F_1, \ldots, F_n]]$. For $1 \leq i \leq n$ let $a_i = F_i(0)$. The jacobian matrices of $(F_1 - a_1, \ldots, F_n - a_n)$ and F coincide. Since the latter is invertible in $K[[x_1, \ldots, x_n]]$, we conclude from the local inversion theorem (see Appendix 2) that

$$K[[x_1, \ldots, x_n]] = K[[F_1 - a_1, \ldots, F_n - a_n]] = K[[F_1, \ldots, F_n]].$$

Thus B is zero on $K[[x_1, \ldots, x_n]]$, as required.

We now return to the Jacobian conjecture. The next theorem is of the type: a conjecture implies a conjecture! To simplify the proof we introduce the following notation. Let $a \in A_n$. The map $\mathbf{ad}_a : A_n \to A_n$ is defined by

$$\mathbf{ad}_a(b) = [a, b].$$

This is a K-linear map, but it is not a K-algebra homomorphism.

4.2 THEOREM. *Let $F : K^n \to K^n$ be a polynomial map and assume that $\Delta F = 1$ everywhere on K^n. If every endomorphism of A_n is an automorphism, then the Jacobian conjecture holds.*

PROOF: Since $\Delta F = 1$, it follows from Lemma 4.1 that $D_1, \ldots, D_n$ are derivations of $K[X]$ which satisfy

$$[D_i, F_j] = D_i(F_j) = \delta_{ij} \text{ and } [D_i, D_j] = 0,$$

for $1 \leq i, j \leq n$. By Ch.1 §3, there exists an endomorphism $\phi : A_n \to A_n$ such that $\phi(x_i) = F_i$ and $\phi(\partial_i) = D_i$, for $1 \leq i \leq n$. Note that for $b \in A_n$,

$$deg(ad_{\partial_i}(b)) = deg[\partial_i, b] \leq deg\, b - 1.$$

Thus given $b \in A_n$, there exists $k \in \mathbb{N}$ such that $(ad_{\partial_i})^k(b) = 0$. Since

$$\phi(ad_{\partial_i}(b)) = ad_{D_i}\phi(b),$$

we have that $(ad_{D_i})^k(\phi(b)) = 0$. Assuming that ϕ is an automorphism, we conclude that D_i is locally nilpotent. It then follows by Proposition 3.1 that $K[F_1, \ldots, F_n] = K[x_1, \ldots, x_n]$, which is the Jacobian conjecture as stated in 2.3.

Once again let us observe that it is not known whether every endomorphism of A_n is an automorphism. This conjecture first appeared in print as 'Problème 11.1' in [Dixmier **68**]. Note, however, that every endomorphism of A_n is injective, by Corollary 2.2.2. Thus to prove the conjecture it is enough to show that every endomorphism of A_n is surjective. Unfortunately this is not known even for $n = 1$.

5. EXERCISES.

5.1 Let $F : \mathbb{R}^2 \to \mathbb{R}^2$ be the polynomial map defined by $F(x, y) = (x^3 + x, y)$. Show that $\Delta F \geq 1$ in $\mathbb{R}^2$, but that it is not constant. Show that F has an inverse but that it is not polynomial.

5.2 Let $g \in K[X] = K[x_1, \ldots, x_n]$. Show that $g(p) = 0$ for every $p \in K^n$ if and only if $g = 0$ as a polynomial.

Hint: Proceed by induction on n. Write g in the form $\sum_0^k g_i x_n^i$, where $g_i \in K[x_1, \ldots, x_{n-1}]$. Suppose that there exists $q \in K^{n-1}$ such that $g_i(q) \neq 0$ for some i. Then

$$g(q, x_n) = \sum_0^k g_i(q) x_n^i$$

is a polynomial in one variable with infinitely many roots.

5.3 Let d be a positive integer. Consider the polynomial map $F : \mathbb{R}^2 \to \mathbb{R}^2$ defined by $F(x, y) = ((x - y)^d + y - 2x, (x - y)^d - x)$. Show that $\Delta F = 1$ everywhere in $\mathbb{R}^2$ and that F has a polynomial inverse.

5.4 Let k be a field of characteristic $p > 0$. Consider the map $F : k^n \to k^n$ defined by

$$F(x_1, \ldots, x_n) = (x_1 + x_1^p, x_2, \ldots, x_n).$$

Show that $J(F)$ is the identity matrix but that F cannot be invertible. Hint: $x_1 + x_1^p$ is not an irreducible polynomial.

5.5 Let $F : \mathbb{C}^n \to \mathbb{C}^n$ be a polynomial map. Suppose that the coordinate functions of F have degree at most 2. Show that if $\Delta F \neq 0$ everywhere in $\mathbb{C}^n$, then

$$F(p) - F(q) = JF(\frac{p+q}{2})(p - q).$$

Use this to prove that F must be injective.

This is enough to prove the Jacobian conjecture for quadratic maps because by [Bass, Connell and Wright 82, Theorem 2.1] an injective polynomial map of $\mathbb{C}^n$ to itself must be bijective.

5.6 Which of the following derivations are locally nilpotent in $K[x_1, x_2]$?

(1) $x_1 \partial_1 + x_2 \partial_2$
(2) $x_1 \partial_1 + \partial_2$

5.7 Check in detail that the D_i defined in §4 are derivations of $K(X)$.

5.8 Let R be a commutative ring and D, D' be derivations of R. Show that $[D, D']$ is a derivation of R.

5.9 Let D be a derivation of $K[x_1, \ldots, x_n]$ that is zero on the constants. Show that:

(1) D can be extended to the power series ring $K[[x_1, \ldots, x_n]]$.
(2) If Δ is a power series such that $\Delta(0) \neq 0$ then $\Delta^{-1} \cdot D$ is a derivation of the power series ring $K[[x_1, \ldots, x_n]]$.

CHAPTER 5

MODULES OVER THE WEYL ALGEBRA

This chapter collects a number of important examples of modules over the Weyl algebra. The prototype of all the examples we discuss here is the polynomial ring in n variables; and with it we shall begin. The reader is expected to be familiar with the basic notions of module theory, as explained in [Cohn 84, Ch.10].

1. The polynomial ring.

In Ch. 1, the Weyl algebra was constructed as a subring of an endomorphism ring. Writing $K[X]$ for the polynomial ring $K[x_1, \ldots, x_n]$ we have that $A_n(K)$ is a subring of $End_K K[X]$. One deduces from this that the polynomial ring is a left A_n-module. Thus the action of x_i on $K[X]$ is by straightforward multiplication; whilst ∂_i acts by differentiation with respect to x_i. This is a very important example, and we shall study it in some detail.

Let us first recall some basic definitions. Let R be a ring. An R-module is *irreducible*, or *simple*, if it has no proper submodules. Let M be a left R-module. An element $u \in M$ is a *torsion element* if $ann_R(u)$ is a non-zero left ideal. If every element of M is torsion, then M is called a *torsion module*.

1.1 Lemma. *Let R be a ring and M an irreducible left R-module.*

(1) *If $0 \neq u \in M$, then $M \cong R/ann_R(u)$.*

(2) *If R is not a division ring, then M is a torsion module.*

PROOF: Consider the map $\phi : R \longrightarrow M$ defined by $\phi(1) = u$. It is a homomorphism of R- modules. Since $u \neq 0$ and M is irreducible, ϕ is surjective. Thus (1) follows from the fact that $\ker \phi = ann_R(u)$.

Now suppose that $ann_R(u) = 0$ for some $0 \neq u \in M$. It follows from (1) that $M \cong R$. Since M is irreducible, this can happen only if the left ideals of R are trivial. But in this case R is a division ring, contradicting the hypothesis. Thus $ann_R u \neq 0$, which proves (2).

Let us apply these results to the A_n-module $K[X]$.

1.2 PROPOSITION. *$K[X]$ is an irreducible, torsion A_n-module. Besides this,*

$$K[X] \cong A_n / \sum_1^n A_n \partial_i.$$

PROOF: First of all 1 is clearly a generator of $K[X]$. Now suppose that $f \neq 0$ is a polynomial and consider the submodule $A_n \cdot f$. Let $x_1^{i_1} \ldots x_n^{i_n}$ be a monomial of maximal possible degree among the monomials that appear in f with non-zero coefficients. Let a be its coefficient. Thus $\partial_1^{i_1} \ldots \partial_n^{i_n} \cdot f = i_1! \ldots i_n! \cdot a$ is a non-zero constant in the submodule generated by f. Hence $A_n \cdot f = K[X]$. Thus $K[X]$ is irreducible. Since A_n is not a division ring, it follows from Lemma 1.1(1) that $K[X]$ is a torsion module.

Now 1 is a non-zero element of $K[X]$ which is annihilated by $\partial_1, \ldots, \partial_n$. Hence the left ideal J generated by $\partial_1, \ldots, \partial_n$ is contained in $ann_{A_n} 1$. Conversely, let $P \in ann_{A_n}(1)$. Then P may be written in the form $f + Q$, where $Q \in J$ and $f \in K[X]$. Thus $0 = P \cdot 1 = f \cdot 1$, which implies that $f = 0$. Therefore $P = Q \in J$. We conclude that $J = ann_{A_n}(1)$. The isomorphism now follows from Lemma 1.1(1).

We may generalize this example as follows. Choose $g_1, \ldots, g_n \in K[X]$ and consider the left ideal J of A_n generated by $\partial_1 - g_1, \ldots, \partial_n - g_n$. Every element of A_n is of the form $f + P$, for $f \in K[X]$ and $P \in J$; see Exercise 4.1. Hence the map $\psi : A_n / J \longrightarrow K[X]$ defined by $\psi(f + J) = f$ is an isomorphism of K-vector spaces. Although the action of the x's is preserved under this isomorphism, it is not an isomorphism of A_n-modules. Indeed, if f is a polynomial, then

$$\partial_i(f + J) = \frac{\partial f}{\partial x_i} + f \cdot \partial_i + J = \frac{\partial f}{\partial x_i} + f \cdot g_i + J.$$

Thus, $\psi(\partial_i \cdot (f + J)) = (\partial_i + g_i) \cdot f$, where the right hand side is to be calculated in $K[X]$ with its natural action. The module A_n / J is irreducible; the proof is similar to that of Proposition 1.2.

Another module that is closely related to $K[X]$ is $A_n / \sum_1^n A_n \cdot x_i$. As a K-vector space it is isomorphic to $K[\partial] = K[\partial_1, \ldots, \partial_n]$, the set of polynomials in $\partial_1, \ldots, \partial_n$. Using this isomorphism, we may identify the action of A_n

directly on $K[\partial]$: the ∂'s act by multiplication, whilst x_i acting on ∂_j gives $-\delta_{ij} \cdot 1$. Apart from the obvious similarities, the modules $K[\partial]$ and $K[X]$ are related in a deeper way that will be explained in the next section.

The emphasis on irreducible modules in this chapter is justified by the fact that they play the rôle of building blocks in module theory. However, it is not true that every A_n-module is a direct sum of irreducible modules. The sense in which we say that (interesting) A_n-modules are built up from irreducible modules is explained in Ch.10.

2. Twisting.

We begin with a general construction. Let R be a ring and M a left R-module. Suppose that σ is an automorphism of R. We shall define a new left module M_σ, as follows. As an abelian group, $M_\sigma = M$. The difference lies in the action of R on M_σ. Let $a \in R$ and $u \in M$, define $a \bullet u = \sigma(a)u$. A routine calculation shows that M_σ is a left R-module. It is called the *twisted module* of M by σ. Not surprisingly, M_σ inherits many of the properties of M.

2.1 PROPOSITION. *Let R be a ring, M a left R-module and σ an automorphism of R. Then:*

(1) *M_σ is irreducible if and only if M is irreducible.*

(2) *M_σ is a torsion module if and only if M is a torsion module.*

(3) *If N is a submodule of M then $(M/N)_\sigma \cong M_\sigma/N_\sigma$.*

(4) *Let J be a left ideal of R. Set $\sigma(J) = \{\sigma(r) : r \in J\}$. Then $\sigma(J)$ is a left ideal of A_n and $(R/J)_\sigma \cong R/\sigma^{-1}(J)$.*

PROOF: An R-module M is irreducible if and only if given any non-zero $u, v \in M$ there exists $a \in R$ such that $au = v$. This equation translates as $\sigma^{-1}(a) \bullet u = v$ in M_σ, which proves (1). Similarly, the equation $au = 0$ becomes $\sigma^{-1}(a) \bullet u = 0$; which proves (2). Now (3) is an immediate application of the first homomorphism theorem.

To prove (4), note that $\sigma(J)$ is a left ideal, since σ is an automorphism of A_n. Let $\phi : R \longrightarrow (R/J)_\sigma$ be the homomorphism of R-modules defined by $\phi(1) = 1 + J$. If $b \in R$, then

$$\phi(b) = b \bullet \phi(1) = \sigma(b) + J.$$

Hence ϕ is surjective, and its kernel is $\sigma^{-1}(J)$. Thus (4) also follows from the first homomorphism theorem.

Let us apply this construction to A_n. An important example is the Fourier transform. Let $\mathcal{F}$ be the automorphism of A_n defined by $\mathcal{F}(x_i) = \partial_i$ and $\mathcal{F}(\partial_i) = -x_i$; see Exercise 1.4.8. Let M be a left A_n-module. The twisted module $M_{\mathcal{F}}$ is called the *Fourier transform* of M. The reason for the name is clear, $\mathcal{F}$ transforms a differential operator with constant coefficients into a polynomial.

2.2 PROPOSITION. *The Fourier transform of $K[X]$ is $K[\partial]$.*

PROOF: It follows from Proposition 1.2 that $K[X] \cong A_n/J$, where $J = \sum_1^n A_n \cdot \partial_i$. Since $\mathcal{F}^{-1}(J) = \sum_1^n A_n \cdot x_i$ we may apply Proposition 2.1(4) to get the desired result.

It follows from Propositions 2.1(1) and 2.2 that $K[\partial]$ is irreducible. The Fourier transform will reappear later on in this book. Let us consider other examples of twisting. For $i = 1, \ldots, n$, let $g_i \in K[x_i]$. Note that $[\partial_i + g_i, \partial_j + g_j] = 0$, for $1 \le i, j \le n$. Hence there exists an automorphism σ of A_n which maps x_i to itself and ∂_i to $\partial_i + g_i$. A straightforward calculation shows that

$$K[X]_\sigma \cong A_n/\sum_1^n (A_n(\partial_i - g_i)),$$

which is a particular case of the example considered in §1. Note that $K[X]_\sigma$ is irreducible, by Proposition 1.2 and Proposition 2.1(1). In fact $K[X]_\sigma$ is irreducible for every automorphism σ of A_n.

We may use the above construction to produce an infinite family of non-isomorphic irreducible left A_n-modules. For every positive integer r let σ_r be the automorphism of A_n which satisfies $\sigma_r(x_i) = x_i$ and $\sigma_r(\partial_i) = \partial_i - x_i^r$.

2.3 THEOREM. *The modules $K[X]_{\sigma_r}$ form an infinite family of pairwise non-isomorphic irreducible modules over A_n.*

PROOF: Let $r < t$, and suppose that there exists an isomorphism, $\phi : K[X]_{\sigma_r} \longrightarrow K[X]_{\sigma_t}$. Since $K[X]_{\sigma_r}$ is irreducible, it is generated by 1. Thus ϕ is completely determined by the image of 1; say $\phi(1) = f \neq 0$. Now the equation $\phi(\partial_i \bullet 1) = \partial_i \bullet \phi(1)$ translates as the differential equation

$$\frac{\partial f}{\partial x_i} = (x_i^t - x_i^r)f.$$

The left hand side of the equation has degree $\leq \deg f - 1$. Since $f \neq 0$ and $r < t$, the right hand side has degree $\deg f + t$. This is a contradiction, so the theorem is proved.

The construction of irreducible A_n-modules leads to many important questions and we shall return to it several times.

3. Holomorphic functions.

Let U be an open subset of $\mathbb{C}$. The set $\mathcal{H}(U)$ of holomorphic functions defined on U contains the polynomial ring $\mathbb{C}[z]$. We will make it into a left $A_1(\mathbb{C})$-module. Denote the generators of $A_1(\mathbb{C})$ by z and $\partial = d/dz$. Then z acts by multiplication, whilst ∂ acts by differentiation on $\mathcal{H}(U)$. It is routine to check that the actions are well-defined and satisfy the required properties; see Appendix 1. As an $A_1(\mathbb{C})$-module, $\mathcal{H}(U)$ is not irreducible. This is the same as saying that there are holomorphic functions which are not polynomials. On the other hand, a torsion element of $\mathcal{H}(U)$ is a holomorphic function which satisfies an ordinary differential equation with polynomial coefficients. These elements exist. For example, $\exp(z)$ is a solution of $d/dz - 1$. However, $\mathcal{H}(U)$ is not a torsion module. This is a more interesting result that we now prove. In fact we show that the function $\exp(\exp(z))$ does not satisfy a polynomial differential equation. First a technical lemma.

3.1 LEMMA. *Let $h(z)$ be the holomorphic function $\exp(\exp(z))$. For every positive integer m there exists a polynomial $F_m(x) \in \mathbb{C}[x]$ of degree m such that*

$$d^m h/dz^m = F_m(e^z)h(z).$$

PROOF: Since $dh/dz = e^z \cdot h(z)$, the result holds for $m = 1$ with $F_1(z) = z$. Suppose, by induction, that the formula holds for m. Using the formula $d^{m+1}h/dz^{m+1} = d(d^m h/dz^m)/dz$ and the induction hypothesis, we have that

$$d^{m+1}h/dz^{m+1} = F_m'(e^z)h(z)e^z + F_m(e^z)F_1(e^z)h(z),$$

where F'_m is the derivative of the polynomial F_m. Putting

$$F_{m+1}(z) = F'_m(z)z + F_m(z)F_1(z)$$

we have the desired formula. Besides this, the degree of F_{m+1} is $\deg F_m + 1 = m + 1$, by the induction hypothesis.

In the proof of the next proposition we use the fact that the exponential function is not algebraic. A complex function $f \in \mathcal{H}(U)$ is *algebraic* if there exists a non-zero polynomial $G(x, y) \in \mathbb{C}[x, y]$ such that $G(z, f(z)) = 0$, for all $z \in U$. For a proof that e^z is not algebraic, see Hardy's delightful book *The integration of functions of a single variable*, [Hardy **28**, Ch.V, §16].

3.2 PROPOSITION. *The function $h(z) = \exp(\exp(z))$ is not a torsion element of the $A_1(\mathbb{C})$-module $\mathcal{H}(U)$.*

PROOF: Let $P = \sum_0^r f_i(z)\partial^i$ be an element of $A_1(\mathbb{C})$. We assume that $f_r \neq 0$. Then $P \cdot h = 0$ in $\mathcal{H}(U)$ is equivalent, by definition, to the differential equation

$$\sum_0^r f_i(z)\frac{d^i h}{dz^i} = 0.$$

Using Lemma 3.1 and the fact that $h(z) = \exp(\exp(z)) \neq 0$ for every $z \in \mathbb{C}$, we obtain an equation of the form

$$\sum_0^r f_i(z)F_i(e^z) = 0.$$

This is equivalent to $G(z, e^z) = 0$, where $G(x, y) = \sum_0^r f_i(x)F_i(y)$ is a polynomial in two variables. Since $F_i(y)$ is a polynomial in y of degree i, the polynomial $G(x, y)$ must be non-zero. This implies that e^z is an algebraic function; a contradiction.

In the next chapter, $\mathcal{H}(U)$ will play the rôle of module of solutions for differential equations represented in $\mathcal{D}$-module language.

4. EXERCISES.

4.1 Let $g_1, \ldots, g_n$ be polynomials in $K[X]$.

(1) Show, by induction on m, that ∂_i^m may be written in the form $D(\partial_i - g_i) + f$, where $D \in A_n$ and $f \in K[X]$.

(2) Conclude that every element of A_n can be put in the form $Q + f$, where $Q \in \sum_1^n A_n(\partial_i - g_i)$ and $f \in K[X]$.

4.2 Show that if $g_i \in K[x_i]$ then the module $A_n / \sum_1^n (A_n(\partial_i - g_i))$ is irreducible.

4.3 A left R-module M is cyclic if $M = R \cdot u$, for some $u \in M$. Show that an irreducible module is always cyclic.

4.4 Let R be a simple ring that is not a division ring. Suppose that M, M' are non-zero left torsion R-modules. Show that if M is cyclic and M' is irreducible, then $M \oplus M'$ is cyclic.

Hint: Let u be a generator of M. Choose $0 \neq a \in R$ such that $au = 0$. Since R is simple $aM' \neq 0$. Let $v \in M'$ with $av \neq 0$. Then $u + v$ generates $M \oplus M'$.

4.5 Using Exercise **4.4**, show that the direct sum of any finite number of irreducible A_n-modules is cyclic.

4.6 Let M, M' be left R-modules and let σ be an automorphism of R. Show that $(M \oplus M')_\sigma \cong M_\sigma \oplus M'_\sigma$.

4.7 Show that $\mathcal{H}(U)$ is not a cyclic $A_1(\mathbb{C})$-module.

Hint: Let $h \in \mathcal{H}(U)$ be a generator. Show that h cannot be constant and that $\exp(h) \notin A_1(\mathbb{C}) \cdot h$.

4.8 Let U be an open set of $\mathbb{R}^n$ and $C^\infty(U)$ be the real vector space of all functions of class C^∞ defined on U. Let x_i act by multiplication and ∂_i by differentiation on $C^\infty(U)$. Show that this makes $C^\infty(U)$ into a left $A_n(\mathbb{R})$-module. Is $C^\infty(U)$ an irreducible $A_n(\mathbb{R})$-module? Is it a torsion module? Is it cyclic?

4.9 Are $\sin(e^z)$ and $\cos(e^z)$ torsion elements of the $A_1(\mathbb{C})$-module $\mathcal{H}(\mathbb{C})$?

4.10 Are $\exp(\sin z)$ and $\exp(\cos z)$ torsion elements of the $A_1(\mathbb{C})$-module $\mathcal{H}(\mathbb{C})$?

4.11 Show that $\cos z$ and $\sin z$ are torsion elements of $\mathcal{H}(\mathbb{C})$.

4.12 Let U be the open set $\mathbb{C} \setminus (-\infty, 0]$. If $\alpha \in \mathbb{R}$, show that z^α is a torsion element of the $A_1(\mathbb{C})$-module $\mathcal{H}(U)$.

CHAPTER 6

DIFFERENTIAL EQUATIONS

We are now ready to justify the statement that the theory of D-modules offers an algebraic approach to linear differential equations. We begin by describing a system of differential equations and its polynomial solutions in module theoretic language. This leads to more general kinds of solutions: distributions, hyperfunctions and microfunctions. We end with a description of the module of microfunctions in dimension 1.

1. The D-module of an equation.

Let P be an operator in A_n. It may be represented in the form $\sum_\alpha g_\alpha \partial^\alpha$, where $\alpha \in \mathbb{N}^n$ and $g_\alpha \in K[x_1, \ldots, x_n] = K[X]$. This differential operator gives rise to the equation

$$P(f) = \sum_\alpha g_\alpha \partial_\alpha(f) = 0,$$

where f may be a polynomial or, if $K = \mathbb{R}$, a C^∞ function on the variables $x_1, \ldots, x_n$. More generally, if $P_1, \ldots, P_m$ are differential operators in A_n, then we have a system of differential equations

(1.1) $$P_1(f) = \cdots = P_m(f) = 0.$$

In this section we want to associate to this system a finitely generated left A_n-module in a canonical way. This will allow us to give a purely algebraic description of the polynomial solutions of (1.1).

The A_n-module *associated* to the system of differential equations (1.1) is $A_n / \sum_1^m A_n P_i$. This definition is justified by the next theorem. A *polynomial solution* of (1.1) is a polynomial $f \in K[X]$ which satisfies $P_i(f) = 0$, for $i = 1, \ldots, m$. The set of all polynomial solutions of (1.1) forms a K-vector space.

1.2 THEOREM. *Let M be the A_n-module associated with the system (1.1).*
The vector space of polynomial solutions of the system (1.1) is isomorphic
to $\text{Hom}_{A_n}(M, K[X])$.

PROOF: Let $f \in K[X]$ be a polynomial solution of (1.1). Consider the
homomorphism $\sigma_f : A_n \longrightarrow K[X]$ which maps $1 \in A_n$ to f. If $Q \in J = \sum_1^m A_n P_i$, then

$$Q(f) = 0.$$

Hence σ_f defines a homomorphism

$$\overline{\sigma_f} : M \longrightarrow K[X].$$

Thus to a polynomial solution f of (1.1) we associate $\overline{\sigma_f} \in \text{Hom}_{A_n}(M, K[X])$.
Furthermore, the map $f \longmapsto \overline{\sigma_f}$ is a linear transformation. Indeed, if $f, g \in K[X]$ and $\lambda \in K$, we have that

$$\overline{\sigma_{f+\lambda g}}(P + J) = P(f + \lambda g) = (\overline{\sigma_f} + \lambda\overline{\sigma_g})(P)$$

by the linearity of the operator P.

The inverse map may be explicitly defined. If $\tau \in \text{Hom}_{A_n}(M, K[X])$, then
it maps the class of 1 in M to a polynomial h. An easy calculation shows
that h is a solution of (1.1) and that $\tau = \sigma_h$. Hence the rule $\tau \longmapsto \tau(1)$
defines the inverse linear transformation.

Be warned that $\text{Hom}_{A_n}(M, K[X])$ is neither an A_n- module nor even a
$K[X]$-module; it is only a K-vector space. Worse still: it may be a vector
space of infinite dimension; see Exercise 4.1.

Although we restricted ourselves to polynomial solutions of the system
(1.1) this is not really necessary. Suppose that $P_1, \ldots, P_m \in A_n(\mathbb{R})$. The set
of C^∞ functions defined on an open set U of $\mathbb{R}^n$, denoted by $C^\infty(U)$, is a left
$A_n(\mathbb{R})$-module; see Exercise 5.4.8. Proceeding as above one shows that the
C^∞-solutions of (1.1) correspond to homomorphisms in $\text{Hom}_{A_n}(M, C^\infty(U))$.

These examples inspire us to make the following definition. Let $\mathcal{S}$ be a
left A_n-module; and let M be a finitely generated left A_n-module. We will
call $\text{Hom}_{A_n}(M, \mathcal{S})$ the *solution space* of M in $\mathcal{S}$. Note that we have taken

care *not* to require that S be finitely generated. For example, $C^\infty(\mathbb{R}^n)$ is not
a finitely generated $A_n(\mathbb{R})$-module; see Exercise 4.4. On the other hand, a
system of differential equations will always correspond to a finitely generated
module. The module associated to (1.1) is even cyclic.

The advantage of a definition of this sort is that it allows us to introduce
generalized solutions of differential equations in a natural way. All one has to
do is choose an appropriate A_n-module S. This includes solutions in terms
of distributions, hyperfunctions and microfunctions. These generalized solutions may be necessary, as some differential equations do not have solutions
in terms of ordinary functions. Here is an example in one variable. Consider
the operator x in $A_1(\mathbb{C})$. The differential equation $xf = 0$ does not have a
non-zero solution even if we require only that f be continuous. However, this
equation has a solution in terms of distributions, the famous Dirac δ-function.

In the next sections we construct the module of microfunctions in one
variable in an algebraic way and express the Dirac δ as a microfunction. It
is then easy to check that $x.\delta = 0$.

2. Direct limit of modules.

The construction of the module of microfunctions makes use of direct lim-
its. This has independent interest and we begin by discussing it in some
detail.

Let $\mathcal{I}$ be a set with a relation $\leq$. We say that $\mathcal{I}$ is *pre-ordered* if $\leq$ is
reflexive and transitive. A pre-ordered set $\mathcal{I}$ is *directed* if, given $i, j \in \mathcal{I}$,
there exists $k \in \mathcal{I}$ such that $k \leq i$ and $k \leq j$. Directed sets will play the rôle
of index sets for our construction.

Let R be a ring and $\mathcal{I}$ be a directed set. Suppose that to every $i \in \mathcal{I}$
we associate a left R-module M_i. We say that $\{M_i : i \in \mathcal{I}\}$ is a *directed
family* of left R-modules if given any $i, j \in \mathcal{I}$, satisfying $i \leq j$, there exists a
homomorphism of R-modules

$$\pi_{ji} : M_j \longrightarrow M_i,$$

and if $i \leq j \leq k$, then

$$\pi_{ji} \cdot \pi_{kj} = \pi_{ki}.$$

An example will make things clearer. Let $D(\epsilon)$ be the open disk of centre 0 and radius ϵ in $\mathbb{C}$. Let $\mathcal{H}(\epsilon) = \mathcal{H}(D(\epsilon))$ be the set of all holomorphic functions defined in $D(\epsilon)$. Recall that $\mathcal{H}(\epsilon)$ is a left A_1-module; see Ch. 5, §3. Take $\mathbb{R}$ to be the index set. If $\epsilon \le \epsilon'$ in $\mathbb{R}$, then $\mathcal{H}(\epsilon') \subseteq \mathcal{H}(\epsilon)$. Hence we may take

$$\pi_{\epsilon'\epsilon} : \mathcal{H}(\epsilon') \longrightarrow \mathcal{H}(\epsilon)$$

to be the restriction of a holomorphic function in $\mathcal{H}(\epsilon')$ to $D(\epsilon)$. This gives us a directed family of A_1-modules.

We return to the general construction. Let $\{M_i : i \in \mathcal{I}\}$ be a directed family of left R-modules. Denote by $\mathcal{U}$ the disjoint union of the modules M_i; it may be identified with the set of all pairs (u, i), where $u \in M_i$. We define an equivalence relation in $\mathcal{U}$ as follows: $(u, i), (v, j) \in U$ are *equivalent* if and only if there exists $k \in \mathcal{I}$ such that $k \le i$, $k \le j$ and $\pi_{ik}(u) = \pi_{jk}(v)$. The *direct limit* of the family $\{M_i : i \in \mathcal{I}\}$, denoted by $\varinjlim M_i$, is the quotient set of $\mathcal{U}$ by this equivalence relation. To simplify the notation, we will write (u, i) for the equivalence class in $\varinjlim M_i$ as well as for its representative in $\mathcal{U}$.

Let us apply this construction to the family $\{\mathcal{H}(\epsilon) : \epsilon \in \mathbb{R}\}$. Things are made a little simpler in this case because $\mathbb{R}$ is completely ordered. An element of $\varinjlim \mathcal{H}(\epsilon)$ is represented by a pair (f, ϵ) where $f \in \mathcal{H}(\epsilon)$. Assuming that $\epsilon \le \epsilon'$, we have that two pairs (f, ϵ) and (g, ϵ') are equal in $\varinjlim \mathcal{H}(\epsilon)$ when $g(z) = f(z)$, for every $z \in D(\epsilon)$. The elements of $\mathcal{H}_0 = \varinjlim \mathcal{H}(\epsilon)$ are called *germs* of holomorphic functions at 0.

We have not yet finished the construction, because we want to turn the direct limit into a module. Once again we return to the general situation. Let $(u, i), (v, j) \in \varinjlim M_i$ and $a \in R$. Choose $k \in \mathcal{I}$ such that $k \le i$ and $k \le j$. The *sum* $(u, i) + (v, j)$ is the element

$$(\pi_{ik}(u) + \pi_{jk}(v), k).$$

The *product* $a \cdot (u, i)$ is (au, i).

We must show that these operations are independent of the various representatives of classes used to define them. Since most of this is routine, we check only that the sum is independent of the choice of k. Suppose that k, k' are both less than or equal to i and j. We want to show that $S = (\pi_{ik}(u) + \pi_{jk}(v), k)$ equals $S' = (\pi_{ik'}(u) + \pi_{jk'}(v), k')$ in $\varinjlim M_i$. Choose $r \in \mathcal{I}$, such that $r \leq k$ and $r \leq k'$. Thus

$$\pi_{kr}(\pi_{ik}(u) + \pi_{jk}(v)) = \pi_{k'r}(\pi_{ik'}(u) + \pi_{jk'}(v))$$

and both equal $\pi_{ir}(u) + \pi_{jr}(v)$. Hence $S = S' = (\pi_{ir}(u) + \pi_{jr}(v), r)$ in $\varinjlim M_i$, as required.

The sum and product by scalar in $\varinjlim M_i$ are defined using the corresponding operations in M_i. Thus the usual properties of the sum and scalar product in a module hold for $\varinjlim M_i$. The details are left to the reader.

In the example $\varinjlim \mathcal{H}(\epsilon)$, the index set $\mathbb{R}$ is totally ordered. Thus $(f, \epsilon) + (g, \epsilon')$ is $(f + g, \epsilon'')$ where $\epsilon'' = min\{\epsilon, \epsilon'\}$. The scalar product is $P \cdot (f, \epsilon) = (P(f), \epsilon)$, for every $P \in A_1(\mathbb{C})$.

3. MICROFUNCTIONS.

The module of microfunctions is constructed as a direct limit. We must first define the family of A_1-modules, the direct limit of which we will be calculating. As in §2, let $D(\epsilon)$ be the open disk of $\mathbb{C}$ of centre 0 and radius ϵ. Let $D'(\epsilon) = D(\epsilon) \setminus 0$.

The *universal cover* of $D'(\epsilon)$ is the set $\tilde{D}(\epsilon) = \{z \in \mathbb{C} : Re(z) < \log(\epsilon)\}$. The projection π of $\tilde{D}(\epsilon)$ on $D'(\epsilon)$ is defined by $\pi(z) = e^z$. Note that π is surjective: if $\rho < \epsilon$ is a positive real number, then $\rho e^{i\theta} = \pi(\log(\rho) + i\theta)$, and $\log(\rho) + i\theta \in \tilde{D}(\epsilon)$. The relative positions of these sets and maps are shown in the following diagram.

$$\tilde{D}(\epsilon)$$
$$\pi \downarrow$$
$$D'(\epsilon) \longrightarrow D(\epsilon)$$

We make the set $\mathcal{H}(\tilde{D}(\epsilon))$ of functions that are holomorphic in $\tilde{D}(\epsilon)$ into an $A_1(\mathbb{C})$-module. Let $h \in \mathcal{H}(\tilde{D}(\epsilon))$. The action of a polynomial $f \in \mathbb{C}[x]$ on h is given by $f \bullet h = f(e^z)h(z)$. The operator $\partial = d/dx$ acts on h by the formula $\partial \bullet h = h'(z)e^{-z}$. A calculation (left to the reader) shows that these actions are well-defined.

3.1 PROPOSITION. *The map*

$$\pi^* : \mathcal{H}(D'(\epsilon)) \longrightarrow \mathcal{H}(\tilde{D}(\epsilon))$$

defined by $\pi^*(h)(z) = h(\pi(z))$ *is an injective homomorphism of* $A_1(\mathbb{C})$-*modules.*

PROOF: Since π is surjective, π^* must be injective. Let us check that π^* is a homomorphism of $A_1(\mathbb{C})$-modules. Suppose that $h \in \mathcal{H}(D'(\epsilon))$. If $f \in \mathbb{C}[x]$, then $\pi^*(fh) = (fh)(e^z)$ and $f \bullet h(e^z) = f(e^z)h(e^z)$ are equal. Similarly,

$$\partial \bullet \pi^*(h) = (h'(e^z)e^z)e^{-z} = h'(e^z)$$

is equal to $\pi^*(\partial \cdot h)$. The proof that π^* preserves the sum is an easy exercise

Since $D'(\epsilon) \subseteq D(\epsilon)$, we have that $\mathcal{H}(D(\epsilon))$ is a submodule of $\mathcal{H}(D'(\epsilon))$. Let $\mathcal{M}_\epsilon$ denote the quotient module $\mathcal{H}(\tilde{D}(\epsilon))/\pi^*(\mathcal{H}(D(\epsilon)))$. If $\epsilon' \leq \epsilon$, then $\tilde{D}(\epsilon') \subseteq \tilde{D}(\epsilon)$. Thus $\mathcal{H}(\tilde{D}(\epsilon)) \subseteq \mathcal{H}(\tilde{D}(\epsilon'))$. This induces a homomorphism of $A_1(\mathbb{C})$-modules

$$\tau_{\epsilon\epsilon'} : \mathcal{M}_\epsilon \longrightarrow \mathcal{M}_{\epsilon'}.$$

Hence $\{\mathcal{M}_\epsilon : \epsilon \in \mathbb{R}\}$ is a directed family of $A_1(\mathbb{C})$-modules, and we may take its direct limit. This limit, denoted by $\mathcal{M}$, is called the module of microfunctions.

The canonical projection of $\mathcal{H}(\tilde{D}(\epsilon))$ onto $\mathcal{M}_\epsilon$ is compatible with the limit, and determines a homomorphism of A_1-modules

$$can : \mathcal{H}(\tilde{D}(\epsilon)) \longrightarrow \mathcal{M}.$$

The function $e^{-z}/2\pi i$ is holomorphic in $\tilde{D}(\epsilon)$. But $e^{-z}/2\pi i$ is also the image of $1/2\pi iz$ under π^*. However, $1/2\pi iz$ is not holomorphic in $D(\epsilon)$. Hence $can(e^{-z}/2\pi i)$ is a non-zero element of $\mathcal{M}$, it is called the *Dirac delta* microfunction, and denoted by δ.

As observed at the end of §1, the equation $xh = 0$ has no analytic (or C^∞) solution, but it is satisfied by the Dirac delta. This is easily checked. Note first that

$$x\delta = can(e^z e^{-z}/2\pi i) = can(1/2\pi i).$$

But $1/2\pi i$, being a constant, is holomorphic in $D(\epsilon)$, hence zero in $\mathcal{M}$. Thus $x\delta = 0$, as required.

Another important example is the *Heaviside* microfunction, defined by $Y = can(z/2\pi i)$. Note that $z/2\pi i$ is the image of $\log(z)/2\pi i$ under π^*. Since $\log(z)/2\pi i$ is not holomorphic in $D(\epsilon)$, the hyperfunction Y is non-zero. Moreover, Y is the integral of δ:

$$\partial \cdot Y = can(e^{-z}/2\pi i) = \delta.$$

Microfunctions can be used to classify certain A_1-modules in terms of *quivers*, a combinatorial object. This is discussed in detail in [Briançon and Maisonobe **84**] and [Malgrange **91**].

4. EXERCISES.

4.1 Show that the K-vector space of polynomial solutions in $K[x_1, x_2]$ of $x_1\partial_2 - x_2\partial_1$ has infinite dimension.

4.2 Show that the set of polynomial solutions in $\mathbb{C}[x]$ of a differential operator of $A_1(\mathbb{C})$ has finite dimension.

4.3 Let U be an open set of $\mathbb{R}^n$. Using Baire's theorem and the fact that $C^\infty(U)$ is a metrizable vector space show that any basis of $C^\infty(U)$ as a real vector space is uncountable.
Hint: Suppose the basis $\{v_i : i \in \mathbb{N}\}$ is countable. Let V_k be the subspace generated by $v_1, \ldots, v_k$. Show that V_k is a closed set of $C^\infty(U)$ with empty interior and that $\bigcup_k V_k = C^\infty(U)$. Obtain a contradiction using Baire.
A proof that $C^\infty(U)$ is a metrizable vector space is found in [Rudin **91**, Ch.1].

4.4 Show that $C^\infty(U)$ is not finitely generated as an $A_n(\mathbb{R})$-module.
Hint: Suppose that it is generated by $f_1, \ldots, f_k$, and show that $x^\alpha\partial^\beta \cdot f_i$ forms a countable set of generators for $C^\infty(U)$ as a real vector space, for $\alpha, \beta \in \mathbb{N}^n$ and $1 \le i \le k$. This contradicts Exercise 4.3.

Exercises 4.5 to 4.7 explain the construction of the module of hyperfunctions. This is very similar to the construction of the microfunctions presented in §3.

4.5 Let Ω be an open interval of $\mathbb{R}$. A *complex neighbourhood* of Ω is an open set U of $\mathbb{C}$ that contains Ω. Consider the set

$$\mathcal{U} = \{U : U \text{ is a complex neighbourhood of } \Omega\}.$$

Show that $\mathcal{U}$ is a directed set for the order $\supseteq$.

4.6 Let Ω be an open interval of $\mathbb{R}$ and U a complex neighbourhood of Ω.

(1) Show that $\mathcal{H}(U) \subseteq \mathcal{H}(U \backslash \Omega)$ and that both are $A_1(\mathbb{R})$-modules, where a polynomial acts by multiplication and ∂ by differentiation.

(2) Show that $\mathcal{H}(U \backslash \Omega)/\mathcal{H}(U)$ is a directed family of $A_1(\mathbb{R})$-modules with respect to the directed set $\mathcal{U}$.

(3) Let $\mathcal{B}(\Omega) = \varinjlim \mathcal{H}(U \backslash \Omega)/\mathcal{H}(U)$ and $h \in \mathcal{H}(U \backslash \Omega)$, for some $U \in \mathcal{U}$. Denote by $[h]$ the image of h in $\mathcal{B}(\Omega)$. Show that if h can be extended to a holomorphic function on U then $[h] = 0$.

(4) Using the notation of the previous item, show that if $f \in \mathbb{R}[x]$ then $f \cdot [h] = [fh]$ and $\partial \cdot [h] = [h']$.

$\mathcal{B}(\Omega)$ is called the *module of hyperfunctions of* Ω.

4.7 Let $\Omega = (0,1)$. The Heaviside hyperfunction is $Y = [\log(-z)/2\pi i]$ and the Dirac hyperfunction $\delta = [1/2\pi i z]$. Show that $\partial \cdot Y = \delta$.

4.8 Show that the submodule of the module of microfunctions $\mathcal{M}$ generated by δ is isomorphic to $\mathbb{C}[\partial]$.

4.9 Let δ' be the first derivative of the Dirac microfunction. Let $A_1(\mathbb{C})\delta'$ be the submodule of $\mathcal{M}$ generated by δ'. Show that

(1) $A_1(\mathbb{C})\delta' = A_1(\mathbb{C})\delta$.

(2) $A_1(\mathbb{C})\delta' \cong A_1(\mathbb{C})/J$, where J is the left ideal of $A_1(\mathbb{C})$ generated by x^2 and $x\partial + 2$.

Hint: If $Q \in A_1(\mathbb{C})$ satisfies $Q\delta' = 0$ then we have $Q \cdot \partial \in A_1(\mathbb{C})x$, the annihilator of δ. Write $Q = Q_2 x^2 + Q_1 x + Q_0$, where $Q_2 \in A_1(\mathbb{C})$, $Q_0, Q_1 \in \mathbb{C}[\partial]$, and calculate $Q \cdot \partial$.

4.10 Let δ^m be the m-th derivative of the Dirac microfunction δ. Show that $A_1(\mathbb{C})\delta^m$ is isomorphic to $A_1(\mathbb{C})/J$, where J is the left ideal of $A_1(\mathbb{C})$ generated by x^m and $x\partial + m$.

Hint: Induction and Exercise 4.9.

CHAPTER 7

GRADED AND FILTERED MODULES

Simple rings are very hard to study because most techniques in ring theory depend on the existence of two-sided ideals. In the case of the Weyl algebra, however, we have a way out. As we saw in Ch. 2, one may define a degree for the elements of the Weyl algebra. Using this degree, we construct a commutative ring, which works as a shadow of A_n. We may then draw an outline of what A_n really looks like. This is the best method we have for understanding the structure of A_n and of its modules.

1. GRADED RINGS

An important feature of a polynomial ring is that it admits a degree function. We want to generalize and formalize what it means for an algebra to have a degree. This leads to the definition of graded rings. These rings find their justification in algebraic geometry, more precisely in projective algebraic geometry; for details see [Hartshorne **77**, Ch. 1, §2]. For the sake of completeness, we define graded rings without assuming commutativity.

Let R be a K-algebra. We say that R is *graded* if there are K-vector subspaces R_i, $i \in \mathbb{N}$, such that

(1) $R = \bigoplus_{i \in \mathbb{N}} R_i$,

(2) $R_i \cdot R_j \subseteq R_{i+j}$.

The R_i are called the *homogeneous components* of R. The elements of R_i are the *homogeneous elements of degree i*. If $R_i = 0$ when $i < 0$ then we say that the grading is *positive*. From now on all graded algebras will have a positive grading unless explicitly stated otherwise.

The most important example of a graded algebra is the polynomial ring $K[x_1, \ldots, x_n]$. The monomials $x_1^{k_1} \ldots x_n^{k_n}$ with $k_1 + \cdots + k_n = m$ form a basis of the homogeneous component of degree m. Noncommutative graded rings have been in evidence recently in the theory of quantum groups and noncommutative geometry. The *quantum plane*, for example, corresponds to the K-algebra generated by two elements x, y which satisfy the relation

$xy = \lambda yx$, for some $\lambda \in K$; see Exercise 6.1. For more details see [Manin 88].

The most important graded rings that come up in algebraic geometry are quotients of polynomial rings. They are constructed as follows. Let R be a graded K-algebra. A two-sided ideal I of R is a *graded ideal* if $I = \bigoplus_{i\geq 0}(I \cap R_i)$. Thus a graded ideal is generated by homogeneous elements. The converse also holds: an ideal generated by homogeneous elements must be graded.

Now let $S = \bigoplus_{i\geq 0} S_i$ be another graded K-algebra. A homomorphism of K-algebras $\phi : R \longrightarrow S$ is *graded* if $\phi(R_i) \subseteq S_i$. Thus a graded homomorphism is one that preserves the degree. The concepts of graded homomorphism and graded ideal are related, as shown in the next result.

1.1 PROPOSITION. *Let $R = \bigoplus_{i\geq 0} R_i$ and $S = \bigoplus_{i\geq 0} S_i$ be graded algebras over K.*

    (1) *The kernel of a graded homomorphism of K-algebras $\phi : R \longrightarrow S$ is a graded two-sided ideal of R.*

    (2) *If I is a graded two-sided ideal of R then R/I is a graded K-algebra.*

PROOF: Suppose first that ϕ is a graded homomorphism. Let $a = a_0 \oplus \cdots \oplus a_s$ be an element of the kernel of ϕ. Thus,

$$\phi(a) = \phi(a_0) + \cdots + \phi(a_s).$$

Since ϕ is graded, this sum is direct. Thus $a_i \in \ker(\phi)$ for $i = 0, \ldots, s$. This proves (1).

Conversely, let $I = \ker(\phi)$ be a graded two-sided ideal. Then we may decompose the quotient ring R/I into a direct sum of K-vector spaces,

$$R/I \cong \bigoplus_{i\geq 0}(R_i/(I \cap R_i)).$$

If $a_i \in R_i$ and $a_j \in R_j$ then $(a_i + I)(a_j + I) = a_i a_j + I$ corresponds to an element of $R_{i+j}/(I \cap R_{i+j})$ under this isomorphism. Hence R/I is a graded ring, which proves (2).

This gives us a way to generate examples of graded rings. Let $F_1, \ldots, F_k$ be homogeneous polynomials in $K[X]$. The quotient $K[X]/(F_1, \ldots, F_k)$ is a graded ring.

A graded algebra admits a special kind of module. Let $R = \bigoplus_{i \geq 0} R_i$ be a graded K-algebra. A left R-module M is a *graded module* if there exist K-vector spaces M_i, for $i \geq 0$, such that

(1) $M = \bigoplus_{i \geq 0} M_i$,

(2) $R_i \cdot M_j \subseteq M_{i+j}$.

The M_i are the *homogeneous components of degree i* of M. Note that the definition of graded module depends on the graded structure chosen for the algebra R.

We may also define graded submodules and graded module homomorphisms, mimicking the corresponding definitions for graded rings. Let R be a graded K-algebra and M, M' be graded left R-modules. A submodule N of M is a *graded submodule* if $N = \bigoplus_{i \geq 0} (N \cap M_i)$. An R-module homomorphism $\theta : M \longrightarrow M'$ is *graded* if $\theta(M_i) \subseteq M_i'$. It follows that $\ker(\theta)$ is a graded submodule and that the quotient module M/N is a graded R-module; the proof is similar to that of Proposition 1.1.

This allows us to produce examples of graded modules. Let $R = \bigoplus_{i \geq 0} R_i$ be a graded K-algebra and R^n be the free left R-module of rank n. This module has a natural grading, its k-th homogeneous component is the vector space

$$\sum_{i_1 + \cdots + i_n = k} (R_{i_1} \oplus \cdots \oplus R_{i_n}).$$

If L is a graded submodule of R^n, then R^n/L is a finitely generated graded left module. The importance of graded rings is explained in the next section.

2. FILTERED RINGS.

In Ch. 2 we introduced the degree of an operator in the Weyl algebra. However, this degree cannot be used to make A_n into a graded ring. The problem is that an element like $\partial_1 x_1$ ought to be homogeneous of degree 2, but it is equal to $x_1 \partial_1 + 1$, which is not homogeneous. To use this degree effectively we must generalize graded rings, to get filtered rings.

Let R be a K-algebra. A family $\mathcal{F} = \{F_i\}_{i \geq 0}$ of K-vector spaces is a *filtration* of R if

(1) $F_0 \subseteq F_1 \subseteq F_2 \subseteq \cdots \subseteq R$,

(2) $R = \bigcup_{i \geq 0} F_i$,

(3) $F_i \cdot F_j \subseteq F_{i+j}$.

If an algebra has a filtration it is called a *filtered algebra*. It is convenient to use the convention that $F_j = \{0\}$ if $j < 0$.

Let us show that every graded algebra is filtered. Suppose that $G = \bigoplus_{i \geq 0} G_i$ is a graded algebra. Consider the vector spaces $F_k = \bigoplus_0^k G_i$. Clearly $F_k \subseteq F_{k+1}$ and their union is the whole of G. Since

$$F_k.F_m = \bigoplus_{i+j \leq k+m} G_i G_j,$$

and $G_i G_j \subseteq G_{i+j}$, we have that $F_k F_m \subseteq F_{k+m}$. Hence $\{F_k\}_{k \geq 0}$ is a filtration of G. On the other hand there are filtered algebras which do not have a natural grading. This happens to the Weyl algebra; which, however, has many different filtrations.

The first filtration of A_n which we will discuss is the *Bernstein filtration*. It is the filtration defined using the degree of operators in A_n. Denote by B_k the set of all operators of A_n of degree $\leq k$. These are vector subspaces of A_n. Conditions (1) and (2) of a filtration are clearly satisfied by the B_k, whilst (3) is a consequence of Theorem 2.1.1(2). Thus $\mathcal{B} = \{B_k\}_{k \in \mathbb{N}}$ is a filtration of A_n. We also write $\mathcal{B}(A_n)$ and $B_k(A_n)$ whenever necessary.

The Bernstein filtration has a very special feature: each B_k is a vector space of finite dimension. A basis for B_k is determined by the monomials $x^\alpha \partial^\beta$ with $|\alpha| + |\beta| \leq k$. In particular, $B_0 = K$ and $\{1, x_1, \ldots x_n, \partial_1, \ldots, \partial_n\}$ is a basis of B_1. The fact that B_k has finite dimension will be fundamental in Ch. 9.

Another important example of a filtration for A_n is the *order filtration*, denoted by $\mathcal{C}$. As in Ch. 3, §2, denote by C_k the vector space of all operators of order $\leq k$ in A_n. It is clear that properties (1) and (2) of a filtration hold for $\mathcal{C}$, and (3) is a consequence of Proposition 3.1.2. Note that $C_0 = K[X]$ is an infinite dimensional K-vector space. Despite this drawback, the order filtration has the advantage that, unlike the Bernstein filtration, it is well-defined for other rings of differential operators.

3. ASSOCIATED GRADED ALGEBRA.

We may use a filtration of an algebra to construct a graded algebra. This is very useful because many properties of this graded algebra pass on to its parent filtered algebra, as we shall see in later chapters. Let R be a K-algebra. Suppose that $\mathcal{F} = \{F_i\}_{i \in \mathbb{N}}$ is a filtration of R. As a first step in the construction of the graded algebra, we introduce the *symbol map of order* k, which is the canonical projection of vector spaces

$$\sigma_k : F_k \longrightarrow F_k/F_{k-1}.$$

Thus for an operator $d \in F_k$, the symbol $\sigma_k(d)$ is non-zero if and only if $d \notin F_{k-1}$.

Consider now the K-vector space

$$gr^{\mathcal{F}} R = \bigoplus_{i \geq 0} (F_i/F_{i-1}).$$

We want to make it into a graded ring. For that it is enough to define the multiplication of two homogeneous elements, and extend it by linearity. A homogeneous element of $gr^{\mathcal{F}} R$ is of the form $\sigma_k(a)$ for some $a \in F_k$. Let $\sigma_m(b)$, for $b \in F_m$ be another homogeneous element, and define their product by

$$\sigma_k(a)\sigma_m(b) = \sigma_{m+k}(ab).$$

A straightforward verification shows that $gr^{\mathcal{F}} R$ with this multiplication is a graded K-algebra, with homogeneous components F_i/F_{i-1}. This is called the *graded algebra of R associated with the filtration $\mathcal{F}$.*

The definition of the multiplication in $gr^{\mathcal{F}} R$ hides some surprises. The best way to illustrate these is with an example, which turns out to be the most important application of this construction to be used in this book. Put $S_n = gr^{\mathcal{B}} A_n$.

3.1 THEOREM. *The graded algebra S_n is isomorphic to the polynomial ring over K in $2n$ variables.*

PROOF: For $i = 1, \ldots, n$, let $y_i = \sigma_1(x_i)$ and $y_{i+n} = \sigma_1(\partial_i)$. We break the proof into several steps.

First step: S_n *is generated by* $y_1, \ldots, y_{2n}$ *as a K-algebra.*

It is enough to prove this for the homogeneous elements of S_n. But a homogeneous element of S_n is of the form $\sigma_k(d)$, for some element d of A_n of degree k. Now d is a linear combination of monomials $x^\alpha \partial^\beta$, with $|\alpha| + |\beta| \le k$. If $|\alpha| + |\beta| = k$, then

$$\sigma_k(x^\alpha \partial^\beta) = (y_1^{\alpha_1} \ldots y_n^{\alpha_n})(y_{n+1}^{\beta_1} \ldots y_{2n}^{\beta_n}).$$

Thus $\sigma_k(d)$ is a linear combination of monomials in $y_1, \ldots, y_{2n}$ of degree k, as we wanted to prove.

Second step: S_n *is a commutative ring.*

Since S_n is generated by $y_1, \ldots, y_{2n}$, we need only show that these elements commute in S_n. For $i = 1, \ldots, n$, we have that $y_i y_{i+n} = \sigma_2(x_i \partial_i)$ and $y_{i+n} y_i = \sigma_2(\partial_i x_i)$. Since $\partial_i x_i = x_i \partial_i + 1$, one has that

$$\sigma_2(\partial_i x_i) = \sigma_2(x_i \partial_i).$$

Thus $y_i y_{i+n} = y_{i+n} y_i$. It is even easier to check that y_i commutes with y_j when $j \neq i + n$, since the corresponding elements in A_n commute.

Let $K[z_1, \ldots, z_{2n}]$ be the ring of polynomials in $2n$ variables. The previous two steps allow us to define a surjective ring homomorphism

$$\phi : K[z_1, \ldots, z_{2n}] \longrightarrow S_n$$

by $\phi(z_i) = y_i$. Since the z's have degree 1 in $K[z_1, \ldots, z_{2n}]$ and the y's have degree 1 in S_n, ϕ is a graded homomorphism of K-algebras.

Third step: ϕ *is injective.*

Let $F \in K[z_1, \ldots, z_{2n}]$, and suppose that $\phi(F) = 0$. Since ϕ is a graded homomorphism, we may assume that F is a homogeneous polynomial. Let

$$F(y_1, \ldots, y_{2n}) = \sum c_{\alpha\beta} y_1^{\alpha_1} \ldots y_n^{\alpha_n} \cdot y_{n+1}^{\beta_1} \ldots y_{2n}^{\beta_n}$$

where $\alpha_1 + \cdots + \alpha_n + \beta_1 + \cdots + \beta_n = k$. Define an operator d of A_n by the formula

$$d = \sum c_{\alpha\beta} x_1^{\alpha_1} \ldots x_n^{\alpha_n} \cdot \partial_1^{\beta_1} \ldots \partial_n^{\beta_n}.$$

Then $\sigma_k(d) = F(y_1, \ldots, y_{2n})$.

If $\sigma_k(d) = \phi(F) = 0$, then $d \in B_{k-1}$. Hence d may be written as linear combination of monomials $x^\alpha \partial^\beta$ with $|\alpha| + |\beta| < k$. By construction, d is also a linear combination of monomials of degree k. Hence, by Proposition 1.2.1, all the coefficients $c_{\alpha\beta}$ above are zero. Thus F is the polynomial zero and ϕ is injective, as required.

One may ask, at this point, what the algebra $gr^{\mathcal{C}} A_n$ associated to the filtration by order looks like. The answer is that it too is isomorphic to a polynomial ring in $2n$ variables, see Exercise 6.5. However, for the rest of the book we will deal almost exclusively with the Bernstein filtration. The advantages of the Bernstein filtration over $\mathcal{C}$ will only be apparent in Ch. 9. Till then most of the results that will be proved for the Bernstein filtration also hold for the filtration by order. Since the proofs are always very similar, we will leave the latter as exercises for the reader.

4. FILTERED MODULES.

To define a filtered module we must start with a filtered ring. For the sake of simplicity we shall give the definitions only for the Weyl algebra with the Bernstein filtration.

Let M be a left A_n-module. A family $\Gamma = \{\Gamma_i\}_{i \geq 0}$ of K-vector spaces of M is a *filtration* of M if it satisfies

(1) $\Gamma_0 \subseteq \Gamma_1 \subseteq \cdots \subseteq M$,
(2) $\bigcup_{i \geq 0} \Gamma_i = M$,
(3) $B_i \Gamma_j \subseteq \Gamma_{i+j}$.

Although this is the standard definition of a module filtration, we shall require that the filtrations used in this book satisfy an additional condition. First note that (3), with $i = 0$, implies that each Γ_j is a K-vector space. The fourth condition that a filtration must satisfy is:

(4) Γ_i is a K-vector space of *finite dimension*.

The convention that $\Gamma_j = \{0\}$ if $j < 0$ remains in force.

Of course $\mathcal{B}$ is a filtration of A_n as an A_n-module. A more interesting example is the A_n-module $K[X]$. The vector spaces Γ_i of all polynomials of degree $\leq i$ form a filtration of $K[X]$ for the Bernstein filtration $\mathcal{B}$.

Following the pattern of §3, we may define the graded module associated with a filtered module. Let M be a left A_n-module M and let Γ be a filtration of M with respect to $\mathcal{B}$. Define the *symbol map of order k* of the filtration Γ to be the canonical projection

$$\mu_k : \Gamma_k \longrightarrow \Gamma_k/\Gamma_{k-1}.$$

Now put

$$gr^\Gamma M = \bigoplus_{i \geq 0}(\Gamma_i/\Gamma_{i-1}).$$

We will define an action of S_n on this vector space. If $a \in F_k$ and $u \in \Gamma_i$ let

$$\sigma_k(a).\mu_i(u) = \mu_{i+k}(au).$$

Extending this formula by linearity we get an action of S_n on $gr^\Gamma M$. We leave it to the reader to check that this action satisfies the required properties. The graded S_n-module $gr^\Gamma M$ is called the *graded module associated to the filtration Γ*.

Let us return to a previous example. Let Γ be the filtration of $K[X]$ with respect to $\mathcal{B}$ defined above. Then Γ_i/Γ_{i-1} is isomorphic to the vector space of all homogeneous polynomials of degree i. Hence $gr^\Gamma M$ is isomorphic to $K[X]$ as a vector space. However we saw in Theorem 3.1 that S_n is isomorphic to the polynomial ring in $2n$ variables $y_1, \ldots, y_{2n}$. We want to determine the action of the y's on a homogeneous polynomial f of degree r, which is to be thought of as an element of Γ_r/Γ_{r-1}. For $i = 1, \ldots, n$, we have that $y_i \cdot f = x_i f$. For $i = n+1, \ldots, 2n$ we must be more careful. Note that $y_i \cdot f$ is, by definition $\mu_r(\partial_i(f))$. But $\partial_i(f)$ is homogeneous of degree $\leq r-1$. Hence $y_i \cdot f = 0$. In particular $ann_{S_n}(gr^\Gamma M)$ is the ideal generated by $y_{n+1}, \ldots, y_{2n}$.

5. INDUCED FILTRATIONS.

Let M be a left A_n-module with a filtration Γ with respect to $\mathcal{B}$. Suppose that N is a submodule of M. We may use Γ to construct filtrations for both N and M/N. These are called the filtrations *induced* by Γ.

To get a filtration for N put $\Gamma' = \{N \cap \Gamma_i\}_{i \geq 0}$. The inclusion $N \subseteq M$ allows us to define injective linear maps:

$$\phi_k : N \cap \Gamma_k / N \cap \Gamma_{k-1} \to \Gamma_k / \Gamma_{k-1}.$$

These may be put together to produce a linear map:

$$\phi : gr^{\Gamma'} N \to gr^{\Gamma} M.$$

A calculation with homogeneous elements of S_n and $gr^{\Gamma} N$ shows that ϕ is a homomorphism of modules. Since the ϕ_k are injective, so is ϕ. We will sometimes write $gr^{\Gamma'} N \subseteq gr^{\Gamma} M$, for short.

Now consider the quotient module M/N. Let Γ'' be the subspace of M/N defined by

$$\Gamma_k'' = \Gamma_k / (\Gamma_k \cap N).$$

A routine calculation shows that Γ'' is a filtration of M/N. Note that

$$\Gamma_k'' / \Gamma_{k-1}'' \cong \Gamma_k / (\Gamma_{k-1} + \Gamma_k \cap N)$$

and let

$$\pi_k : \Gamma_k / \Gamma_{k-1} \longrightarrow \Gamma_k'' / \Gamma_{k-1}''$$

be the canonical projection. Putting these together we get a K-linear map,

$$\pi : gr^{\Gamma} M \longrightarrow gr^{\Gamma''} M/N.$$

One easily checks that this map is a surjective homomorphism of S_n-modules.

5.1 LEMMA. *Let M be an A_n-module with a filtration Γ compatible with $\mathcal{B}$. The sequence of S_n-modules*

$$0 \to gr^{\Gamma'} N \xrightarrow{\phi} gr^{\Gamma} M \xrightarrow{\pi} gr^{\Gamma''} M/N \to 0$$

is exact.

PROOF: Note that the kernel of the map π_k defined above is

$$(\Gamma_{k-1} + \Gamma_k \cap N) / \Gamma_{k-1} \cong (\Gamma_k \cap N) / (\Gamma_{k-1} \cap N).$$

Thus we have an exact sequence of vector spaces

$$0 \to (\Gamma_k \cap N)/(\Gamma_{k-1} \cap N) \xrightarrow{\phi_k} \Gamma_k/\Gamma_{k-1} \xrightarrow{\pi_k} \Gamma_k/(\Gamma_{k-1} + \Gamma_k \cap N) \to 0.$$

The sequence of S_n-modules in the lemma is obtained by adding these vector space sequences for $k \geq 0$. Hence it is exact.

The exact sequence of Lemma 5.1 is very useful. Here is a typical application. Let d be an operator in A_n of degree r and put $M = A_n/A_n d$. Take $\mathcal{B}$ to be the filtration of A_n as a left A_n-module. The induced filtration in $A_n d$ is $\mathcal{B}'_k = B_{k-r}d$. Thus,

$$\mathcal{B}'_k/\mathcal{B}'_{k-1} = B_{k-r}d/B_{k-r-1}d \cong (B_{k-r}/B_{k-r-1})\,\sigma_r(d).$$

Since B_k/B_{k-1} is the homogeneous component of degree k of S_n, then

$$gr^{\mathcal{B}'}(A_n d) \cong S_n \sigma_r(d).$$

By Lemma 5.1, there is an exact sequence,

$$0 \longrightarrow S_n \sigma_r(d) \longrightarrow S_n \longrightarrow gr^{\mathcal{B}'} M \longrightarrow 0.$$

Therefore, $gr^{\mathcal{B}'} M \cong S_n/S_n \sigma_r(d).$

6. EXERCISES

6.1 Let $F = K\{x, y\}$ be the free algebra in two generators, and I the two-sided ideal of F generated by the relation $xy - \lambda yx$. Show that the quotient ring F/I is a graded ring.

6.2 Let $R = \bigoplus_{i \geq 0} R_i$ be a graded ring and $M = \bigoplus_{i \geq 0} M_i$ a graded R-module. Let $M(k)_i = M_{i-k}$.

(1) Show that $M(k) = \bigoplus_{i \geq 0} M(k)_i$ is a graded R-module.
(2) Show that if $k > 0$ then the identity map $M \longrightarrow M(k)$ is an isomorphism of R-modules, but not a graded isomorphism.

6.3 In this exercise we define a grading for A_1 that is *not positive*. Denote by x and ∂ the generators of A_1. Define $G_k = \{d \in A_1 : [x\partial, d] = kd\}$. Let $K[x\partial]$ be the polynomial ring in the operator $x\partial$. Show that

(1) $G_k = K[x\partial]x^k$ for $k \geq 0$,

(2) $G_k = K[x\partial]\partial^{-k}$ for $k \leq 0$,

(3) $A_1 = \bigoplus_{i \in \mathbb{Z}} G_k$ is a graded ring.

6.4 Let $d \in A_n$. Define the *principal symbol* of d by $\sigma(d) = \sigma_k(d)$, where k is the degree of d and σ_k denotes the symbol map of degree k relative to the Bernstein filtration. Find the principal symbol of the following operators of A_3:

(1) $\partial_1^4 x_1^6 + x_3^2$;

(2) $x_1^6 \partial_2^4 + x_2^4 x_3^3 \partial_1^2 \partial_3 + x_2 x_4 \partial_1^2 + x_1^5$;

(3) $\partial_2^7 + x_3^7 + x_1^4 x_2^3 + \partial_2 x_3^6 + x_2$.

6.5 Let τ_k be the symbol map of order k with respect to the *order* filtration of A_n. Let $\xi_i = \tau_1(\partial_i)$. Show that $gr^{\mathcal{O}} A_n$ is isomorphic to the polynomial ring $K[X][\xi_1, \ldots, \xi_n]$.

6.6 Let R be a filtered K-algebra with a filtration $\mathcal{F}$. Show that if $gr^{\mathcal{F}} R$ is a domain, then so is R.

6.7 Let J be a left ideal of A_n. Define the *symbol ideal* of J to be the ideal $gr(J) = \sum_{k \geq 0} \sigma_k(J \cap B_k)$ of S_n. Let $M = A_n/J$.

(1) Let $\mathcal{B}'$ be the filtration of J induced by the Bernstein filtration. Show that $gr^{\mathcal{B}'} J \cong gr(J)$.

(2) Show that if $\mathcal{B}''$ is the filtration of M induced by the Bernstein filtration, then $gr^{\mathcal{B}''} M \cong S_n/gr(J)$.

6.8 The definition of a filtered A_n-module for the *order filtration* is analogous to that given in §4, except that (4) must be replaced by: M_i is a finitely generated $K[X]$-module, for $i \geq 0$. Find a filtration of the A_n-module $K[X]$ with respect to the order filtration and calculate its associated graded module.

6.9 Let V be the K-vector subspace of A_n with basis $\{x_1, \ldots, x_n, \partial_1, \ldots, \partial_n\}$. Let $Sp(V)$ be the symplectic group on V.

(1) Show that an element $\sigma \in Sp(V)$ can be extended to an automorphism of A_n that preserves the Bernstein filtration.

(2) Show that this automorphism induces an automorphism of $S_n = gr^B A_n$.

6.10 Let $(\lambda_1, \ldots, \lambda_{2n}) \in K^{2n}$. Show that the formulae

$$\sigma(x_i) = x_i - \lambda_i$$

and

$$\sigma(\partial_i) = \partial_i - \lambda_{n+i}$$

define an automorphism of $A_n(K)$ that preserves the Bernstein filtration and induces the identity automorphism on S_n.

6.11 Let $d \in A_n(\mathbb{C})$ be an element of degree 2. Show that there exist an automorphism α of $A_n(\mathbb{C})$ and constants $\lambda_1, \ldots, \lambda_n, r_1, \ldots, r_n$ such that

$$\alpha(d) = \sum_1^n \lambda_i(x_i^2 + \partial_i^2) + r_i.$$

Hint: Use Exercises 6.9 and 6.10, and the fact that quadratic forms are diagonalizable.

CHAPTER 8
NOETHERIAN RINGS AND MODULES

There is very little that one can say about a general ring and its modules. In practice an interesting structure theory will result either if the ring has a topology (which is compatible with its operations), or if it has finite dimension, or some generalization thereof. As an example of the former, we have the theory of C^*-algebras. The latter class includes many important rings: algebras that are finite dimensional over a field, PI rings, artinian rings and noetherian rings. It is the last ones that we now study. In particular, we prove that the Weyl algebra is a noetherian ring.

1. NOETHERIAN MODULES.

In this book we shall be concerned almost exclusively with finitely generated modules. One easily checks that a homomorphic image of a finitely generated module is finitely generated. However a finitely generated module can have a submodule that is not itself finitely generated. An example is the polynomial ring in infinitely many variables $K[x_1, x_2, \ldots]$. Taken as a module over itself this ring is a cyclic left module: it is generated by 1. However, the ideal generated by all the variables $x_1, x_2, \ldots$ cannot be finitely generated: every finite set of polynomials in $K[x_1, x_2, \ldots]$ uses up only finitely many of the variables.

We get around this problem with a definition. A left R-module is called *noetherian* if all its submodules are finitely generated. Examples are easy to come by: vector spaces over K are noetherian K-modules. Every ideal of the polynomial ring in one variable $K[x]$ is a noetherian $K[x]$-module.

There are several equivalent ways to define noetherianness. We chose the most natural. Here are two more.

1.1 THEOREM. *Let M be a left R-module. The following conditions are equivalent:*

(1) *M is noetherian.*

(2) *For every infinite ascending chain $N_1 \subseteq N_2 \subseteq \ldots$ of submodules of M there exists $k \geq 0$ such that $N_i = N_k$ for every $i \geq k$.*

(3) *Every set S of submodules of M has a submodule L which is not properly contained in any submodule of S.*

Condition (2) is known as the *ascending chain condition*. This is probably the most common way in which noetherian modules are defined. Condition (3) is the *maximal condition*: the submodule L is called a *maximal element* of S.

PROOF: Suppose that (1) holds. If $N_1 \subseteq N_2 \subseteq \ldots$ is an infinite ascending chain of submodules of M, then $Q = \bigcup_{i \geq 1} N_i$ is a submodule of M. Thus Q is generated by finitely many elements, say $u_1, \ldots, u_t$. Hence there exists $k \geq 0$ such that $u_1, \ldots, u_t \in N_k$. Thus $Q = N_k = N_i$, for every $i \geq k$, as claimed in (2).

Assume now that (2) holds. We prove (3) by reaching a contradiction. Suppose that S does not contain a maximal element. If $N_1 \subseteq N_2 \subseteq \cdots \subseteq N_k$ is any chain of elements of S, then we can make it longer. Since no element of S is maximal, there must exist N_{k+1} in S such that $N_k \subset N_{k+1}$. In this way we construct a proper infinite ascending chain of submodules of M, thus contradicting (2).

Finally, assume (3) and let N be a submodule of M. Let S be the set of all finitely generated submodules of N, and let L be a maximal element of S. Suppose $L \subset N$, and choose $u \in N \setminus L$. Thus $L + Ru$ is finitely generated and contains L properly: a contradiction. Hence $L = N$, is finitely generated.

In the next proposition we collect the basic properties of noetherian modules that are used in later sections. First, a technical lemma.

1.2 LEMMA. *Let M be a left module over a ring R. Let N, P_1 and P_2 be submodules of M such that $P_2 \subseteq P_1$. If $N + P_1 = N + P_2$ and $N \cap P_1 = N \cap P_2$, then $P_1 = P_2$.*

PROOF: We need only show that $P_1 \subseteq P_2$. Suppose that $u_1 \in P_1$. Since

$$u_1 \in N + P_1 = N + P_2,$$

we have that $u_1 = x + u_2$, for $x \in N$ and $u_2 \in P_2$. Thus

$$x = u_1 - u_2 \in P_1 \cap N = P_2 \cap N.$$

In particular, $x \in P_2$. Hence $u_1 = x + u_2 \in P_2$, as required.

1.3 PROPOSITION. *Let M be a left module over a ring R, and let N be a submodule of M.*

(1) *M is noetherian if and only if M/N and N are noetherian.*

(2) *Let N' be another submodule of M and suppose that $M = N + N'$. If N,N' are noetherian, then so is M.*

PROOF: It is clear that a submodule of a noetherian module is itself noetherian. On the other hand, a submodule of M/N is of the form L/N for some submodule L of M which contains N. If M is noetherian, then L is finitely generated. Thus L/N is finitely generated, and M/N is noetherian.

Conversely, suppose that N and M/N are noetherian. Let $L_1 \subseteq L_2 \subseteq \ldots$ be an infinite ascending chain of submodules of M. Then

$$(L_1 \cap N) \subseteq (L_2 \cap N) \subseteq \ldots$$

is an ascending chain of submodules of N. Since N is noetherian, this chain must stop. In other words, there exists s such that $(L_s \cap N) = (L_i \cap N)$ for all $i \geq s$. Similarly, the chain

$$N + L_1/N \subseteq N + L_2/N \subseteq \ldots$$

stops; that is there exists r such that $N + L_r = N + L_i$, for $i \geq r$. Taking $t = \max\{s, r\}$, we have that $N + L_t = N + L_i$ and $L_t \cap N = L_i \cap N$, for $i \geq t$. Thus by Lemma 1.2, $L_t = L_i$ for $i \geq t$. Hence M is a noetherian ring.

To prove (2), note that $M/N \cong N'/(N' \cap N)$ is noetherian by (1). Since N is noetherian, we may use (1) again to conclude that M is noetherian.

We have pitifully few examples of noetherian modules, but this situation will be rectified in the next section.

2. NOETHERIAN RINGS.

We say that a ring R is a *left noetherian ring* if R is noetherian as a left R-module. Examples are fields and any principal ideal domain, like

$\mathbb{Z}$ or $K[x]$. Actually, the most important rings in algebraic geometry are noetherian, because the ring of polynomials in finitely many variables is noetherian. This was proved by D. Hilbert in 1890 and it is the keystone of commutative algebra. It is known as *Hilbert's basis theorem.* We shall now prove a modernized version of Hilbert's result from which the original theorem follows as an easy consequence.

2.1 THEOREM. *Let R be a commutative noetherian ring. The polynomial ring $R[x]$ is noetherian.*

PROOF: Suppose that $R[x]$ is not noetherian. Let I be an ideal of $R[x]$ that is not finitely generated. We shall construct, inductively, an infinite ascending chain of ideals in R; thereby achieving a contradiction. Choose $f_1 \in I$ of smallest possible degree. Assume, by induction, that $f_1, \ldots, f_k$ have been chosen. Let f_{k+1} be the polynomial of smallest possible degree in $I \setminus (f_1, \ldots, f_k)$. Since I is not finitely generated, this construction produces an infinite sequence of polynomials $f_1, f_2, \ldots \in I$. Let n_i be the degree and a_i the leading coefficient of f_i.

Since R is noetherian, the ascending chain of ideals

$$(a_1) \subseteq (a_1, a_2) \subseteq \ldots$$

must be stationary; say $(a_1, \ldots, a_k) = (a_1, \ldots, a_{k+1})$. Hence $a_{k+1} = \sum_1^k b_i a_i$, for some $b_i \in R$. Put

$$g = f_{k+1} - \sum_1^k b_i x^{n_{k+1} - n_i} f_i.$$

Since, by construction, the degrees satisfy $n_1 \le n_2 \le \ldots$ it follows that g is indeed a polynomial. Note that $g \in I$, but $g \notin (f_1, \ldots, f_k)$. However g has smaller degree than f_{k+1}, a contradiction. Thus I must be finitely generated.

A simple induction using Theorem 2.1 is all that is needed to prove the original basis theorem of Hilbert.

2.2 COROLLARY. *Let K be a field. The polynomial ring $K[x_1, \ldots, x_n]$ is noetherian.*

Although every ideal of $K[x_1, \ldots, x_n]$ is finitely generated, it is not true that there is an upper bound on the number of generators of ideals in this ring. For example, it is easy to construct ideals I_k of $K[x_1, x_2]$ which cannot be generated by less than k elements; see Exercise 4.2.

We shall now use Hilbert's basis theorem to prove that A_n is a left noetherian ring. This will follow from a slightly more general theorem. Recall that $\mathcal{B}$ is the Bernstein filtration of A_n and $S_n = gr^{\mathcal{B}} A_n$.

2.3 THEOREM. *Let M be a left A_n-module with a filtration Γ with respect to the Bernstein filtration $\mathcal{B}$. If $gr^{\Gamma} M$ is a noetherian S_n-module, then M is noetherian.*

PROOF: Let N be a submodule of M, and Γ' the filtration of N induced by Γ; see Ch.7, §5. Since $gr^{\Gamma'} N \subseteq gr^{\Gamma} M$ and the latter is noetherian, we conclude that $gr^{\Gamma'} N$ is finitely generated.

Since the generators of $gr^{\Gamma'} N$ are finite in number, they have degree $\leq m$, for some integer m. We wish to show that N is generated by the elements in Γ'_m. Suppose that it is not, and let k be the smallest integer for which there exists $v \in \Gamma'_k$ which cannot be generated by the elements of Γ'_m. Clearly $k > m$. Let μ_k be the symbol map of order k of Γ'. By the hypothesis on $gr^{\Gamma'} N$, there exist $a_i \in A_n$ and $u_i \in \Gamma'_{r_i}$ such that

$$\mu_k(v) = \sum_1^s \sigma_{k-r_i}(a_i) \mu_{r_i}(u_i)$$

where $r_i \leq m$, for $i = 1, 2, \ldots, s$. Hence

$$v - \sum_1^s a_i u_i \in \Gamma'_{k-1},$$

which, by the minimality of k, may be written as an A_n-linear combination of elements in Γ'_m. Thus v itself may be written as an A_n-linear combination of elements in Γ'_m.

To finish the proof we must show that a finite number of elements of Γ'_m are enough to generate N. But Γ'_m is a subspace of the finite dimensional vector space Γ_m. Hence it has a finite basis, which will be a set of generators for N as A_n-module.

2.4 COROLLARY. *A_n is a left noetherian ring.*

PROOF: The graded ring S_n associated to the Bernstein filtration is a polynomial ring in $2n$ variables by Theorem 7.3.1. Hence, by Hilbert's basis theorem, S_n is noetherian. Thus A_n is left noetherian by Theorem 2.3.

It has already been pointed out in Ch.2 that every left ideal of A_n can be generated by two elements. Therefore, there is an upper bound on the number of generators of left ideals in A_n, which is not the case for a polynomial ring. However, this result is very hard to prove and is beyond our means in this book.

There are good reasons why one ought to rejoice that A_n is a left noetherian ring. For example, it follows from Corollary 2.4 and the next proposition that every finitely generated A_n-module is noetherian.

2.5 PROPOSITION. *Let R be a left noetherian ring. Finitely generated left R-modules are noetherian.*

PROOF: Let M be a finitely generated left R-module. If M is generated by k elements then there exists a surjective homomorphism $\phi : R^k \longrightarrow M$. Since R is left noetherian, it follows from Proposition 1.3(2) that R^k is noetherian. Hence M is noetherian by 1.3(1).

There is really nothing special about left modules in this context, and all the results that we have stated so far hold for right modules with the obvious changes. Thus the ring A_n is also right noetherian, and we shall use this in later chapters, without further comment.

3. GOOD FILTRATIONS.

Let M be a left A_n-module and Γ a filtration of M with respect to the Bernstein filtration $\mathcal{B}$. If $gr^\Gamma M$ is finitely generated, then it is noetherian by Proposition 2.5. Hence M is finitely generated, by Theorem 2.3. However, it is not always true that if M is finitely generated over A_n then $gr^\Gamma M$ is finitely generated over S_n. When $gr^\Gamma M$ is finitely generated we say that Γ is a *good* filtration of M.

It is nonetheless true that every finitely generated A_n-module admits a good filtration. Indeed, if M is generated by $u_1, \ldots, u_s$ then the filtration

Γ of M defined by $\Gamma_k = \sum_1^s B_k u_i$ is good. The graded module $gr^\Gamma M$ is generated over S_n by the symbols of $u_1, \ldots, u_s$.

We need an example of a filtration that is not good. Before that, however, we shall establish an easy criterion to check whether a filtration is good.

3.1 PROPOSITION. *Let M be a left A_n-module. A filtration Γ of M with respect to $\mathcal{B}$ is good if and only if there exists k_0 such that $\Gamma_{i+k} = B_i \Gamma_k$, for all $k \geq k_0$.*

PROOF: Suppose that there exists k_0 such that $\Gamma_{i+k} = B_i \Gamma_k$, for all $k \geq k_0$. The K-vector space Γ_{k_0} has a finite basis. The symbols of the elements in this basis generate $gr^\Gamma M$. Thus Γ is good.

Conversely, suppose that $gr^\Gamma M$ is finitely generated over S_n. Let $u_1, \ldots, u_s$ be elements of M whose symbols generate $gr^\Gamma M$. Assume that $u_j \in \Gamma_{k_j} \setminus \Gamma_{k_j-1}$, for $j = 1, 2, \ldots, s$, and that $k_0 = \max\{k_1, \ldots, k_s\}$.

Let $k \geq k_0$. We prove that $\Gamma_{i+k} = B_i \Gamma_k$ by induction on i. If $i = 0$, the result is obviously true. Suppose that the equality holds for $i - 1$. Pick $v \in \Gamma_{i+k}$. Since $gr^\Gamma M$ is finitely generated, and denoting by μ_k the symbol of order k of Γ, one has

$$\mu_{i+k}(v) \in \sum_{j=1}^s \sigma_{k+i-k_j}(B_{k+i-k_j}) \mu_{k_j}(u_j).$$

Since $B_{i+k-k_j} = B_i.B_{k-k_j}$, we conclude that

$$v \in \sum_{j=1}^s B_i.B_{k-k_j} u_j + \Gamma_{i+k-1}.$$

Now $B_{k-k_j} u_j \in \Gamma_k$ for every j and, by the induction hypothesis, $\Gamma_{k+i-1} = B_{i-1} \Gamma_k$. Since $B_{i-1} \subseteq B_i$, then $v \in B_i \Gamma_k$. Thus $\Gamma_{i+k} \subseteq B_i \Gamma_k$. The other inclusion is obvious; hence we have an equality, as desired.

It is now easy to check whether a filtration is good or not. For example, let Ω be the filtration of the left module A_n defined by $\Omega_k = B_{2k}$. Then we have that $B_i \Omega_k = B_{i+2k}$ is properly contained in $B_{2(i+k)} = \Omega_{i+k}$. By Proposition 3.1, Ω is not a good filtration of A_n with respect to $\mathcal{B}$.

We end this section with a proposition that will allow us to compare two good filtrations.

3.2 PROPOSITION. *Let M be a left A_n-module. Suppose that Γ and Ω are two filtrations of M with respect to $\mathcal{B}$.*

(1) *If Γ is good then there exists k_1 such that $\Gamma_j \subseteq \Omega_{j+k_1}$.*

(2) *If Γ and Ω are good then there exists k_2 such that $\Omega_{j-k_2} \subseteq \Gamma_j \subseteq \Omega_{j+k_2}$.*

PROOF: By Proposition 3.1 there exists k_0 such that $B_i \Gamma_j = \Gamma_{i+j}$, for all $j \geq k_0$ and $i \geq 0$. Since Γ_{k_0} is a finite dimensional vector space over K, there exists k_1 such that $\Gamma_{k_0} \subseteq \Omega_{k_1}$. Thus $\Gamma_{j+k_0} = B_j \Gamma_{k_0}$ is contained in $B_j . \Omega_{k_1} \subseteq \Omega_{j+k_1}$. Hence

$$\Gamma_j \subseteq \Gamma_{j+k_0} \subseteq \Omega_{j+k_1}$$

for all $j \geq 0$, which proves (1). To prove (2) apply (1) twice, swapping Γ and Ω. ∎

The full power of good filtrations will only be felt in the next chapter, where they will be used to define a dimension for finitely generated A_n-modules.

4. EXERCISES.

4.1 A K-algebra R is *affine* if there exist elements $r_1, \ldots, r_s \in R$ so that the monomials $r_1^{m_1} \ldots r_s^{m_s}$ form a K-basis for R.

(1) Show that if R is an affine commutative K-algebra then it is a homomorphic image of a polynomial ring in finitely many variables over K.

(2) Use (1) and Hilbert's basis theorem to conclude that a commutative affine K-algebra is noetherian.

(3) Show that (2) is false if the commutative hypothesis is dropped.

Hint: Construct an infinite ascending chain of left ideals for the free algebra $K\{x, y\}$.

4.2 Let I_k be the ideal of the polynomial ring $K[x_1, x_2]$ generated by the monomials of degree k. In other words, the generators of I_k are of the form $x_1^i x_2^j$ with $i + j = k$. Let $\mathcal{M} = I_1$.

(1) Show that $\mathcal{M}$ is a maximal ideal of $K[x_1, x_2]$, and that $K[x_1, x_2]/\mathcal{M} \cong K$.

(2) Show that $I_k/\mathcal{M}I_k$ is a K-vector space of dimension $k + 1$.

(3) Show that I_k cannot be generated by less than $k + 1$ elements.

4.3 Show that the results of §3 remain true if we replace the Bernstein filtration $\mathcal{B}$ by the filtration by order.

Hint: In some of the proofs it will be necessary to use Hilbert's basis theorem and Proposition 2.5.

4.4 Let M be a left A_n-module with a filtration Γ with respect to $\mathcal{B}$. Suppose that $u_i \in \Gamma_{k_i}$, for $i = 1, \ldots, s$. Show that if $gr^\Gamma M$ is generated by $\mu_{k_i}(u_i)$, for $i = 1, \ldots s$, then M is generated by $u_1, \ldots, u_s$.

4.5 The purpose of this exercise is to show that the converse of Exercise 4.4 is false. Let J be the left ideal of A_3 generated by ∂_1 and $\partial_2 + x_1 \partial_3$. Recall that by Exercise 7.6.7, if $\mathcal{B}'$ is the filtration of J induced by the Bernstein filtration of A_3, then $gr^{\mathcal{B}'} J \cong grJ$.

  (1) Using Exercise 2.4.2, show that grJ is generated by $\sigma_1(\partial_i)$ for $i = 1, 2, 3$.
  (2) Show that although ∂_1 and $\partial_2 + x_1 \partial_3$ generate J, their symbols do not generate grJ.

4.6 Let M be a left A_n-module with a good filtration Γ. Show that the annihilator of $gr^\Gamma M$ is a homogeneous ideal of S_n.

CHAPTER 9

DIMENSION AND MULTIPLICITY

Using the filtered and graded methods of the previous chapters we shall define a dimension for A_n-modules. This is a very useful invariant and it comes naturally associated with another invariant: the multiplicity. The first section contains a result in commutative algebra that is the key to the definition of dimension in §2.

1. THE HILBERT POLYNOMIAL.

Without further ado we state the main theorem of this section. It is expected that the reader will feel somewhat surprised by its content.

1.1 THEOREM. *Let* $M = \bigoplus_{i \geq 0} M_i$ *be a finitely generated graded module over the polynomial ring* $K[x_1, \ldots, x_n]$. *There exist a polynomial* $\chi(t) \in \mathbb{Q}[t]$ *and a positive integer* N *such that*

$$\sum_0^s \dim_K(M_i) = \chi(s)$$

for every $s \geq N$.

The polynomial $\chi(t)$ is known as the *Hilbert polynomial* of M. We need a technical lemma before we come to the proof of the theorem. The following definitions and notations make the proofs more bearable. A *numerical polynomial* is a polynomial $p(t)$ of $\mathbb{Q}[t]$ such that $p(n) \in \mathbb{Z}$ for all integers $n \gg 0$. The *difference function* of a function $f : \mathbb{C} \longrightarrow \mathbb{C}$ is defined by $\Delta f(z) = f(z+1) - f(z)$. Finally, set

$$\binom{t}{r} = t(t-1)\ldots(t-r+1)/r!$$

where r is a positive integer and t is a variable. If $r = 0$ we put $\binom{t}{0} = 1$.

1.2 LEMMA. *The following are true:*

(1) $\Delta\binom{t}{r} = \binom{t}{r-1}$.

(2) *Let* $p(t) \in \mathbb{Q}[t]$ *be a numerical polynomial. There exist integers* $c_0, \ldots, c_k$ *such that*

$$p(t) = \sum_0^k c_{k-i}\binom{t}{i}.$$

   In particular $p(n) \in \mathbb{Z}$, *for all* $n \in \mathbb{Z}$.

(3) *Let* $f : \mathbb{Z} \longrightarrow \mathbb{Z}$ *be a function. Suppose that there exists a numerical polynomial* $q(t) \in \mathbb{Q}[t]$ *such that* $\Delta f(n) = q(n)$, *for all* $n \gg 0$. *Then there exists a numerical polynomial* $p(t) \in \mathbb{Q}[t]$ *such that* $f(n) = p(n)$, *for all* $n \gg 0$.

PROOF: (1) follows from a straightforward calculation that we omit . Let us prove (2) by induction on the degree of $p(t)$. If $p(t)$ has degree 0, then it equals an integer. Suppose that the result holds for all numerical polynomials of degree $\leq k - 1$. Let $p(t)$ be a numerical polynomial of degree k. Since the leading term of $\binom{t}{r}$ is $t^r/r!$, there exist rational numbers $c_0, \ldots, c_k$ such that $p(t) = \sum_0^k c_{k-i}\binom{t}{i}$. To complete the proof we must show that $c_i \in \mathbb{Z}$. By (1),

$$\Delta p(t) = \sum_1^k c_{k-i}\binom{t}{i-1}.$$

Since $\Delta p(t)$ has degree $k - 1$, it follows by induction that $c_{k-1}, \ldots, c_0 \in \mathbb{Z}$. Thus $p(t) - c_k$ is integer valued. Since $p(t)$ is a numerical polynomial, it follows that for $r \gg 0$ both $p(r)$ and $p(r) - c_k$ are integers. Hence $c_k \in \mathbb{Z}$, and we have proved (2).

We now prove (3). By (2), the numerical polynomial $q(t)$ may be written in the form

$$q(t) = \sum_0^k c_{k-i}\binom{t}{i},$$

with $c_0, \ldots, c_k \in \mathbb{Z}$. Let

$$p(t) = \sum_0^k c_{k-i}\binom{t}{i+1}.$$

By (1), we have that $\Delta p(t) = q(t)$. Hence $\Delta(f - p)(n) = 0$ for all $n \gg 0$. But this is possible only if $(f - p)(n)$ equals a constant integer, to be called c_{k+1}, for all $n \gg 0$. Thus $f(n) = p(n) + c_{k+1}$, for $n \gg 0$, as required.

PROOF OF THE THEOREM:

The proof is by induction on the number n of variables. If $n = 0$, we are reduced to the base field K. In this case the finitely generated module M is a vector space of finite dimension. Hence it has only finitely many homogeneous components. Thus

$$\sum_{0}^{s} \dim_K M_i = \dim_K M,$$

for $s \gg 0$, and we may choose $\chi(t) = \dim_K M$, a constant.

Suppose, by induction, that the theorem holds for any finitely generated graded module over $K[x_1, \ldots, x_{n-1}]$. Let $M = \bigoplus_{i \geq 0} M_i$ be a finitely generated graded module over $K[x_1, \ldots, x_n]$. Define a function $f : \mathbb{Z} \longrightarrow \mathbb{Z}$ by

$$f(s) = \sum_{-\infty}^{s} \dim_K M_i$$

where $M_i = 0$ if $i < 0$. Let $\phi_i : M_i \longrightarrow M_{i+1}$ be the linear map of vector spaces defined by multiplication by x_n. Set $Q_i = \ker(\phi_i)$ and $L_i = \operatorname{coker}(\phi_i)$. Consider the following exact sequence of vector spaces:

(1.3) $0 \to Q_{i-1} \to M_{i-1} \xrightarrow{\phi_{i-1}} M_i \to L_i \to 0.$

Now $Q = \bigoplus_{i \geq 0} Q_i$ and $L = \bigoplus_{i \geq 0} L_i$ are finitely generated graded modules over $K[x_1, \ldots, x_n]$. They are respectively the kernel and cokernel of the endomorphism of M defined by multiplication by x_n. Moreover, the elements of Q and L are annihilated by x_n, hence these are in fact finitely generated graded $K[x_1, \ldots, x_{n-1}]$-modules. Thus, by induction, there exist polynomials $\chi_1(t), \chi_2(t) \in \mathbb{Q}[t]$ such that

$$\chi_1(s) = \sum_{0}^{s} \dim_K Q_i \quad \text{and} \quad \chi_2(s) = \sum_{0}^{s} \dim_K L_i,$$

for $s \gg 0$.

On the other hand, the dimensions of the vector spaces in (1.3) are related by the formula

$$\dim_K Q_{i-1} - \dim_K M_{i-1} + \dim_K M_i - \dim_K L_i = 0.$$

Adding these up for $0 \le i \le s$ and $s \gg 0$, one obtains

$$\chi_1(s-1) + \Delta f(s) - \chi_2(s) = 0.$$

Hence $\Delta f(s)$ is a numerical polynomial for $s \gg 0$. We may now conclude from Lemma 1.2(3) that $f(s)$ itself is a numerical polynomial for all $s \gg 0$, as required.

2. Dimension and Multiplicity.

We are ready to define the dimension of a module over A_n. Note that the Bernstein filtration $\mathcal{B}$ is essential for the construction that follows.

Let M be a finitely generated left A_n-module. Suppose that Γ is a good filtration of M with respect to the Bernstein filtration $\mathcal{B}$. Denote by $\chi(t, \Gamma, M)$ the Hilbert polynomial of the graded module $gr^\Gamma M$ over the polynomial ring S_n. By Theorem 1.1, we have, for $t \gg 0$,

$$(2.1) \qquad \chi(t, \Gamma, M) = \sum_0^t \dim_K (\Gamma_i/\Gamma_{i-1}) = \dim_K(\Gamma_t).$$

The last equality in (2.1) follows from the fact that $\dim_K$ is additive over exact sequences of vector spaces.

The *dimension* $d(M)$ of M is the degree of $\chi(t, \Gamma, M)$. Let $a_{d(M)}$ be the leading coefficient of $\chi(t, \Gamma, M)$. The *multiplicity* of M is $m(M) = d! a_{d(M)}$. Both numbers are non-negative integers. This is obvious for $d(M)$, and follows from Lemma 1.2(2) for $m(M)$.

The definitions of dimension and multiplicity apparently depend on the good filtration Γ from which the Hilbert polynomial is calculated. Let us show that any choice of good filtration will give the same result. Suppose that Γ and Ω are two good filtrations of M. By Proposition 8.3.2, there exists

k, such that $\Omega_{j-k} \subseteq \Gamma_j \subseteq \Omega_{j+k}$. In particular, $\dim_K \Omega_{j-k} \leq \dim_K \Gamma_j \leq \dim_K \Omega_{j+k}$. We conclude from Theorem 1.1 that for $j \gg 0$

$$\chi(j-k, \Omega, M) \leq \chi(j, \Gamma, M) \leq \chi(j+k, \Omega, M).$$

Since the behaviour of a polynomial at ∞ is determined by its leading term, it follows that $\chi(t, \Omega, M)$ and $\chi(t, \Gamma, M)$ have the same degrees and leading coefficients. Hence $d(M)$ and $m(M)$ are independent of the choice of good filtration for M.

Let us consider a few examples. First let M be the left A_n-module A_n. The Bernstein filtration $\mathcal{B}$ is a good filtration of M and it is possible to calculate $\chi(t, \mathcal{B}, M)$ explicitly in this case. By (2.1) we must determine the dimension of B_t. But the monomials $x^\alpha \partial^\beta$ with $|\alpha| + |\beta| \leq t$ form a basis of B_t as a K-vector space. So it is enough to count the elements of this basis. To do this we must count the non-negative solutions of the equation $\alpha_1 + \cdots + \alpha_n + \beta_1 + \cdots + \beta_n \leq t$. It is an easy exercise in combinatorics to show that there are $\binom{t+2n}{2n}$ such solutions. Hence $\chi(t, \mathcal{B}, M) = \binom{t+2n}{2n}$. As a polynomial in t it has degree $2n$ and leading coefficient $1/(2n)!$. Thus $d(A_n) = 2n$ and $m(A_n) = 1$.

Another A_n-module that we know very well is $K[X] = K[x_1, \ldots, x_n]$. In Ch. 7 we defined a filtration Γ of $K[X]$ such that Γ_t is the space of all polynomials of degree $\leq t$. Since B_i contains the polynomials in $x_1, \ldots, x_n$ of degree i, we have that $B_i \Gamma_t = \Gamma_{t+i}$. Hence Γ is a good filtration. It is easy to show that $\dim_K \Gamma_t = \binom{n+t}{n}$. Thus $\chi(t, \Gamma, M) = \binom{n+t}{n}$ is a polynomial of degree n and leading coefficient $1/n!$. Hence $d(K[X]) = n$ and $m(K[X]) = 1$.

A large class of examples is obtained by twisting a module by an automorphism, as in Ch. 5, §2. Curiously, this will lead us to a different definition for the dimension. Recall that if σ is an automorphism of A_n and M is a left A_n-module, then the action of $a \in A_n$ on $u \in M_\sigma$ is defined by $a \bullet u = \sigma(a)u$.

We begin with the Fourier transform of a module. In this case the automorphism $\mathcal{F}$ is defined by

$$\mathcal{F}(x_i) = \partial_i \text{ and } \mathcal{F}(\partial_i) = -x_i,$$

for $1 \leq i \leq n$. This automorphism $\mathcal{F}$ preserves the Bernstein filtration; thus $\mathcal{F}(B_i) = B_i$. This is very convenient.

2.2 PROPOSITION. *Let M be a finitely generated left A_n-module. Then M and $M_{\mathcal{F}}$ have the same dimension and multiplicity.*

PROOF: Write Γ_0 for a K-vector space whose basis is a set of generators of M. Then $\Gamma_k = B_k \Gamma_0$ is a good filtration for M. Now $M_{\mathcal{F}}$ is also generated by Γ_0, so we may construct a good filtration for $M_{\mathcal{F}}$ by setting $\Omega_k = B_k \bullet \Gamma_0$. Since $\mathcal{F}$ preserves B_k, we have that $\Omega_k = \Gamma_k$. Hence $M_{\mathcal{F}}$ and M have the same Hilbert polynomial, therefore also the same dimension and multiplicity.

Things are a lot more complicated if the automorphism does not preserve the filtration. To get around the problem we must give a different definition for the dimension. Start by choosing a finite number of elements which generate A_n as a K-algebra and let V be the K-vector space generated by these elements and by 1. Put

$$U_0 = K \text{ and } U_k = V^k.$$

This is a filtration of A_n which satisfies $\dim_K U_k < \infty$. Note that if $V = B_1$, then $U_k = B_k$ is the Bernstein filtration of A_n.

Now let M be a finitely generated left A_n-module, with a good filtration Γ with respect to the Bernstein filtration. Without loss of generality we may assume that $\Gamma_k = B_k \Gamma_0$, for $k \geq 0$. Put $\Omega_k = U_k \Gamma_0$ and

$$\delta(M, V) = \inf\{\nu : \dim_K \Omega_s \leq s^\nu \text{ for } s \gg 0\}.$$

2.3 PROPOSITION. *Let V be a vector space whose basis is a finite set of generators for A_n. Then $\delta(M, V) = d(M)$.*

PROOF: Since V is finite dimensional, there exists $r \in \mathbb{N}$ such that $V \subseteq B_r$. Thus $U_k \subseteq B_{rk}$. Furthermore,

$$\Omega_k = U_k \Gamma_0 \subseteq B_{rk} \Gamma_0 \subseteq \Gamma_{rk}.$$

Hence $\dim_K \Omega_k \leq \dim_K \Gamma_{rk}$. In particular, $\delta(M, V) \leq \delta(M, B_1)$. The opposite inequality is proved similarly. Thus $\delta(M, V) = \delta(M, B_1)$. But $\dim_K \Gamma_s = \chi(s, M, \Gamma)$ for $s \gg 0$. Since $d(M)$ is the degree of the polynomial $\chi(s, M, \Gamma)$, then $\delta(M, B_1) = d(M)$, and the proof is complete.

The result we want is a simple corollary of this proposition.

2.4 COROLLARY. *Let M be a finitely generated left A_n-module and σ an automorphism of A_n, then $d(M_\sigma) = d(M)$.*

PROOF: Let $V = \sigma(B_1)$. Since σ is an automorphism, we have that V has generators of A_n for its basis. Define U_k as above. Let Γ be a good filtration of M with respect to the Bernstein filtration, and assume that $\Gamma_k = B_k\Gamma_0$. Now let

$$\Omega_k = B_k \bullet \Gamma_0.$$

This a good filtration for M_σ, hence by Proposition 2.3, $d(M_\sigma) = \delta(M_\sigma, B_1)$. But $\Omega_k = \sigma(B_k)\Gamma_0$. Since $\sigma(B_k) = U_k$, we have that $\Omega_k = U_k\Gamma_0$ and so $\delta(M_\sigma, B_1) = \delta(M, V)$. By Proposition 2.3, the latter equals $d(M)$.

In ring theory $\delta(M, V)$ is called the *Gelfand-Kirillov dimension* of a module. It is discussed in detail, including many applications, in [Krause and Lenagan 85] and [McConnell and Robson 87, Ch. 8].

3. BASIC PROPERTIES.

Let M be a finitely generated left A_n-module and Γ a good filtration of M with respect to $\mathcal{B}$. Let N be a submodule of M. Denote by Γ' and Γ'' the filtrations induced by Γ in N and M/N, respectively; see Ch. 7, §5. It follows from Lemma 7.5.1 that we have an exact sequence of S_n-modules, namely

$$0 \to gr^{\Gamma'}N \to gr^\Gamma M \to gr^{\Gamma''}M/N \to 0.$$

Since Γ is good, $gr^\Gamma M$ is finitely generated. But S_n is a noetherian ring. Hence $gr^{\Gamma'}N$ and $gr^{\Gamma''}(M/N)$ are also finitely generated. Therefore both Γ' and Γ'' are good filtrations.

On the other hand, since the sequence of vector spaces

$$0 \to \Gamma'_k/\Gamma'_{k-1} \to \Gamma_k/\Gamma_{k-1} \to \Gamma''_k/\Gamma''_{k-1} \to 0$$

is exact, we have that

$$\dim_K \Gamma'_k/\Gamma'_{k-1} + dim\Gamma''_k/\Gamma''_{k-1} = \dim_K \Gamma_k/\Gamma_{k-1}.$$

Summing these terms for $k = 0, 1, \ldots, s$ and assuming that $s \gg 0$ one obtains

(3.1) $\chi(s, \Gamma', N) + \chi(s, \Gamma'', M/N) = \chi(s, \Gamma, M).$

3.2 THEOREM. *Let M be a finitely generated left A_n-module and N a submodule of M.*

(1) $d(M) = \max\{d(N), d(M/N)\}$.

(2) *If* $d(N) = d(M/N)$ *then* $m(M) = m(N) + m(M/N)$.

PROOF: Since M is finitely generated it admits a good filtration Γ. Let Γ' and Γ'' be the good filtrations of N and M/N induced by Γ. Thus (3.1) holds for an infinite number of values of s. Hence we have an equality of polynomials:

$$\chi(t, \Gamma', N) + \chi(t, \Gamma'', M/N) = \chi(t, \Gamma, M).$$

Since $d(M)$ corresponds to the degree of $\chi(t, \Gamma, M)$, we have that

$$d(M) \leq \max\{d(N), d(M/N)\}.$$

But the leading coefficients of these polynomials are positive. Thus we must have equality in the formula above, and (1) is proved.

Now, if $d(M/N) = d(N)$ then all these polynomials have the same degree. Thus the leading term of $\chi(t, \Gamma, M)$ is the sum of the leading terms of $\chi(t, \Gamma', N)$ and $\chi(t, \Gamma'', M/N)$, and (2) follows immediately.

This theorem is useful in calculating the dimension of some modules. We saw in §2 that $d(A_n) = 2n$. We may use the theorem to calculate the dimension and multiplicity of a free module of finite rank r: it has dimension $2n$ and multiplicity r. This follows from the following result.

3.3 COROLLARY. *Let $M_1, \ldots, M_k$ be finitely generated left A_n-modules, and $M = M_1 \oplus \cdots \oplus M_k$.*

(1) $d(M) = \max\{d(M_1), \ldots, d(M_k)\}$.

(2) *If* $d(M) = d(M_i)$ *for* $1 \leq i \leq k$, *then* $m(M) = \sum_1^k m(M_i)$.

PROOF: The proof follows by induction if we apply Theorem 3.2 to the exact sequence

$$0 \to M_k \to M \to M_1 \oplus \cdots \oplus M_{k-1} \to 0.$$

We may also use the theorem to get an upper bound on the dimension of a finitely generated A_n-module.

3.4 COROLLARY. *Let M be a finitely generated A_n-module. Then $d(M) \leq 2n$.*

PROOF: Suppose that M is generated by r elements. Then there exists a surjective homomorphism $\phi : A_n^r \to M$. It follows from the theorem that $d(A_n^r) = \max\{d(M), d(\ker \phi)\}$. Since $d(A_n^r) = 2n$ by Corollary 3.3, we conclude that $d(M) \leq 2n$.

This upper bound may be sharpened if the module is a quotient of A_n by a left ideal.

3.5 COROLLARY. *Let I be a non-zero left ideal of A_n. Then $d(A_n/I) \leq 2n - 1$.*

PROOF: First consider the case of a cyclic left ideal. Let $d \in A_n$, and put $I = A_n d$. Then we have an exact sequence

$$0 \to A_n \xrightarrow{\theta} A_n \to A_n/A_n d \to 0$$

where the map θ is defined by $\theta(a) = ad$, for every $a \in A_n$. Suppose, by contradiction, that $d(A_n/A_n d) = 2n$. Then by Theorem 3.2(2), we have that

$$m(A_n) = m(A_n) + m(A_n/A_n d).$$

Since $m(A_n) = 1$ and the multiplicity is a positive number, this equation is impossible. Hence $d(A_n/A_n d) \leq 2n - 1$.

Now for the general case. Let I be a non-zero left ideal of A_n and choose $0 \neq d \in I$. Since $A_n d \subseteq I$, we have that A_n/I is a quotient of $A_n/A_n d$. Since the latter has dimension $\leq 2n - 1$, so does A_n/I, by Theorem 3.2(1).

4. BERNSTEIN'S INEQUALITY.

At the end of the previous section we obtained an upper bound on the dimension of a finitely generated A_n-module. In this section we establish a lower bound. We begin with a lemma from linear algebra.

4.1 LEMMA. *Let M be a finitely generated left A_n-module with filtration Γ with respect to $\mathcal{B}$. Suppose that $\Gamma_0 \neq 0$. The K-linear transformation*

$$\phi : B_i \to \operatorname{Hom}_K(\Gamma_i, \Gamma_{2i})$$

which maps $a \in B_i$ to the linear transformation $\phi_a(u) = au$ is injective.

PROOF: The statement of the lemma is equivalent to $a\Gamma_i \neq 0$ whenever $0 \neq a \in B_i$. We will prove this by induction on i. If $i = 0$, then $B_0 = K$ and the statement above is equivalent to $\Gamma_0 \neq 0$. This is true by hypothesis.

Suppose that if $0 \neq b \in B_{i-1}$ then $b\Gamma_{i-1} \neq 0$. Let a be a non-zero element of B_i. If $a\Gamma_i = 0$, then $a \notin K$. Hence the canonical form of a has a term $cx^\alpha \partial^\beta$ with $c \in K \setminus \{0\}$ and $|\alpha| + |\beta| > 0$. Suppose that $\alpha_i \neq 0$ for this monomial. Then $\alpha_i cx^{\alpha - e_i} \partial^\beta$ is a summand in the canonical form of $[a, \partial_i]$. Thus $[a, \partial_i]$ is a non-zero element of B_{i-1}. Since $a\Gamma_i = 0$, we conclude that

$$[a, \partial_i]\Gamma_{i-1} \subseteq a\partial_i\Gamma_{i-1}.$$

But $\partial_i\Gamma_{i-1} \subseteq \Gamma_i$, hence $[a, \partial_i]\Gamma_{i-1} = 0$, which contradicts the induction hypothesis. We may reach a similar contradiction by assuming that $\beta_i \neq 0$. Thus the lemma is proved.

4.2 THEOREM. *If M is a finitely generated non-zero left A_n-module, then $d(M) \geq n$.*

PROOF: Choose a set of generators for M and let Γ be the good filtration obtained by giving each of these generators degree zero, see Ch. 8, §3. Then $\Gamma_0 \neq 0$. Let $\chi(t) = \chi(t, \Gamma, M)$ be the corresponding Hilbert polynomial.

By the lemma, B_i can be embedded in $\text{Hom}_K(\Gamma_i, \Gamma_{2i})$. In particular,

$$\dim_K B_i \leq \dim_K(\text{Hom}_K(\Gamma_i, \Gamma_{2i})).$$

But $\text{Hom}_K(\Gamma_i, \Gamma_{2i})$ has dimension $\dim_K \Gamma_i \cdot \dim_K \Gamma_{2i}$. Thus, assuming that $i \gg 0$, we get that $\dim_K B_i \leq \chi(i)\chi(2i)$.

On the other hand, $\dim_K B_i = \binom{i+2n}{2n}$ is a polynomial in i of degree $2n$. Hence, as a polynomial in i, $\chi(i)\chi(2i)$ must have degree $\geq 2n$. But the degree of $\chi(i)\chi(2i)$ is $2d(M)$. Thus $d(M) \geq n$, as required.

This inequality was first proved by I. N.Bernstein in [Bernstein 72]; and is often called the *Bernstein inequality*. The proof above is due to A. Joseph. We have already seen that both bounds for the dimension are attained. For example, $d(A_n) = 2n$ and $d(K[X]) = n$. In fact there exist A_n-modules of dimension k for every k in the interval n to $2n$; see Exercise 5.3. The modules of minimal dimension are so important that Ch. 10 will be devoted to them.

5. EXERCISES

5.1 Show that if M is a finitely generated *torsion* A_n-module, then $d(M) \leq 2n - 1$.

5.2 Show that if d is a non-zero operator of A_n then $d(A_n/A_n d) = 2n - 1$. Hint: Let k be the degree of d. There is an exact sequence,

$$0 \to S_n \sigma_k(d) \to S_n \to gr^{\mathcal{B}''}(A_n/A_n d) \to 0,$$

where $\mathcal{B}''$ is the filtration of $A_n/A_n d$ induced by $\mathcal{B}$. Using this sequence show that the Hilbert polynomial of $A_n/A_n d$ is $\binom{2n+t}{2n} - \binom{2n+t-k}{2n-k}$, and that it has degree exactly $2n - 1$.

5.3 Let k be a positive integer such that $1 \leq k \leq n$. Denote by R_k the subalgebra of A_n generated by $x_1, \ldots, x_n$ and $\partial_1, \ldots, \partial_k$. Let $J = A_n \partial_{k+1} + \cdots + A_n \partial_n$.

    (1) Show that the complex vector space $B_i + J/J$ is isomorphic to $B_i \cap R_k$.

    (2) Show that the Hilbert polynomial of A_n/J with respect to the filtration induced by $\mathcal{B}$ is $\binom{n+k+t}{n+k}$.

    (3) Show that A_n/J has dimension $n + k$. What is its multiplicity ?

5.4 Let D be a (noncommutative) domain. We say that D satisfies the Öre condition if $Da \cap Db \neq 0$, for any two non-zero elements $a, b \in D$. A division ring Q is a *left quotient ring* of D if

    (1) D is a subring of Q;

    (2) every non-zero element of D is invertible in Q;

    (3) every element of Q is of the form $b^{-1}a$ where $a, b \in D$ and $b \neq 0$.

Show that if a domain has a left quotient ring then it satisfies the Öre condition. The converse is also true, but it is harder to prove. For details see [Cohn 79, Theorem 12.1.2].

5.5 Show that A_n satisfies the Öre condition.

Hint: Let a, b be non-zero elements of A_n. If $A_n a \cap A_n b = 0$, then A_n maps injectively to $A_n/A_n a \oplus A_n/A_n b$. This contradicts Corollary 3.5.

5.6 Let $d \in A_n(\mathbb{C})$ be an operator of degree 2. Set $M = A_n/A_n d$.

(1) Show that there exists a submodule N of M such that $d(M/N) = n$.

(2) Show that M cannot be an irreducible module if $n \geq 2$.

Hint: By Exercise 7.6.11, there exists an automorphism α of $A_n(\mathbb{C})$ such that $\alpha(d) = \sum_1^n d_i$ where $d_i = \lambda_i(x_i^2 + \partial_i^2) + r_i$ for $\lambda_i, r_i \in \mathbb{C}$. Let J be the left ideal generated by $d_1, \ldots, d_n$. Then A_n/J is a homomorphic image of $M_{\alpha^{-1}} \cong A_n/A_n d$ and $d(A_n/J) = n$. For an easy method to find the dimension of A_n/J see Exercise 13.5.8.

CHAPTER 10

HOLONOMIC MODULES

The most important A_n-modules are the holonomic modules, also known among PDE theorists as maximally overdetermined systems. An A_n-module is holonomic if it has dimension n. Ordinary differential equations with polynomial coefficients correspond to holonomic modules. In this chapter we begin the study of holonomic modules, which will be one of the central topics of the second half of the book.

1. DEFINITION AND EXAMPLES.

A finitely generated left A_n-module is *holonomic* if it is zero, or if it has dimension n. Recall that by Bernstein's inequality this is the minimal possible dimension for a non-zero A_n-module. We already know an example of a holonomic A_n-module, viz. $K[X] = K[x_1, \ldots, x_n]$. We also know that A_n itself is not a holonomic module: it has dimension $2n$.

It is easy to construct holonomic modules if $n = 1$. Let $I \neq 0$ be a left ideal of A_1. By Corollary 9.3.5, $d(A_1/I) \leq 1$. If $I \neq A_1$, then, by Bernstein's inequality, $d(A_1/I) = 1$. Hence A_1/I is a holonomic A_1-module.

This is wonderful source of examples, which will pour forth with the help of the next proposition.

1.1 PROPOSITION. *Let n be a positive integer.*

   (1) *Submodules and quotients of holonomic A_n-modules are holonomic.*

   (2) *Finite sums of holonomic A_n-modules are holonomic.*

PROOF: (1) These follow from Bernstein's inequality. Let M be a left A_n-module, and N a submodule of M. From Theorem 9.3.2, $d(N) \leq d(M)$ and $d(M/N) \leq d(M)$. Since $d(M) = n$, and using Bernstein's inequality, we deduce that $d(N) = d(M/N)$ are also equal to n. Thus N and M/N are holonomic. Now (2) follows from Corollary 9.3.3 and (1).

1.2 COROLLARY. *Finitely generated torsion A_1-modules are holonomic.*

PROOF: Let M be a finitely generated torsion A_1-module. Suppose that it is generated by $u_1, \ldots, u_r$. Since M is torsion, for each $i = 1, 2, \ldots, r$ there exists $0 \neq b_i \in A_1$ such that $b_i u_i = 0$. Hence $A_1 u_i$ is a quotient of $A_1/A_1 b_i$ which is a holonomic module. Thus each $A_1 u_i$ is holonomic. Since M is the sum of $A_1 u_i$, for $i = 1, 2, \ldots, r$, it is holonomic by Proposition 1.1(2).

It is easy to construct torsion A_1-modules. For example, if $0 \neq I \subseteq J$ are left ideals of A_1, then the quotient I/J is a torsion A_1-module. A good exercise, which the reader may wish to try, is to prove that I/J is holonomic using only Proposition 1.1(1). Still on the subject of torsion modules, we have the following proposition.

1.3 PROPOSITION. *Holonomic A_n-modules are torsion modules.*

PROOF: Let M be a holonomic left A_n-module. Suppose that $0 \neq u$ is an element of M. Consider the map $\phi : A_n \to M$ defined by $\phi(a) = au$. Since $\operatorname{im}\phi \subseteq M$, it follows that $d(\operatorname{im}\phi) = n$. Thus by Theorem 9.3.2,

$$2n = d(A_n) = d(\ker \phi).$$

In particular $\ker \phi \neq 0$, and u is a torsion element of M.

Putting 1.2 and 1.3 together, we conclude that for finitely generated A_1-modules, torsion and holonomic are equivalent concepts. However, it is not true that all torsion A_n-modules are holonomic when $n \geq 2$. For example, consider the module $M = A_n/A_n \partial_n$. The Bernstein filtration induces in M a filtration defined by

$$\Gamma_i = B_i/(B_i \cap A_n \partial_n).$$

Note that Γ_i is isomorphic to the vector space generated by the monomials in $x_1, \ldots, x_n$ and $\partial_1, \ldots, \partial_{n-1}$ of degree $\leq i$. We may use this to calculate the Hilbert polynomial of M explicitly, using combinatorics. It equals $\binom{2n+i-1}{2n-1}$. As a polynomial in i, this binomial number has degree $2n - 1$; see Exercises 9.5.2 and 9.5.3. Hence $d(M) = 2n - 1 > n$, if $n \geq 2$. On the other hand, it is very easy to check that M is a torsion module: its elements are annihilated by powers of ∂_n.

2. Basic properties.

Many interesting properties of holonomic modules follow from the fact that they are artinian. A module M over a ring R is said to be *artinian* if given a descending sequence $N_1 \supseteq N_2 \supseteq \ldots$ of submodules of M, there exists k such that $N_k = N_j$, for every $j \geq k$. In other words, every descending chain of submodules of M is stationary. This is like the noetherian property. The following proposition collects some of the basic properties of artinian modules.

2.1 THEOREM. *Let M be a left module over a ring R and let N be a submodule of M.*

(1) *M is artinian if and only if every set S of submodules of M has an element (a minimal element) which does not contain any other element of S.*

(2) *M is artinian if and only if N and M/N are artinian.*

(3) *Let N' be another submodule of M, and suppose that $M = N + N'$. If N, N' are artinian, then so is M.*

The proof of (1) is like that of Theorem 8.1.1 and the proofs of (2) and (3) are like those in Proposition 8.1.3. The details are left to the reader. Let us now see why we are interested in artinian modules.

2.2 THEOREM. *Holonomic modules are artinian.*

PROOF: Let M be a holonomic left A_n-module. Suppose that M has a descending chain of submodules

$$M = N_0 \supseteq N_1 \supseteq N_2 \supseteq \cdots \supseteq N_r.$$

By Theorem 9.3.2 and Proposition 1.1(1), it follows that $m(N_i) = m(N_{i+1}) + m(N_i/N_{i+1})$. Putting these together, we get that

$$m(M) = \sum_0^{r-1} m(N_i/N_{i+1}) + m(N_r) \geq r.$$

Hence M cannot have a descending chain of more than r submodules. In particular M cannot have an infinite descending chain.

Let us dispel any hopes that the reader may have formed, by saying that not all A_n-modules are artinian. For example, A_n is not artinian as a module over itself. It is easy to construct an infinite descending chain; take for instance

$$A_n x_n \supseteq A_n x_n^2 \supseteq A_n x_n^3 \supseteq \dots .$$

More examples of non-artinian A_n-modules are found in Exercises 4.2 and 4.3. A ring R that is artinian as a left R-module is called a *left artinian ring*. The argument above shows that A_n is not a left artinian ring. Artinian rings have a very nice representation theory; see [Reiten **85**].

We return, for a moment, to the general situation. Let M be a left non-zero module over a ring R. Assume that M is both artinian and noetherian. Suppose we have constructed a chain $N_0 \subseteq N_1 \subseteq \dots \subseteq N_k$ of submodules of M such that N_i/N_{i-1} is irreducible. It follows from Theorem 2.1(1) that there exists a submodule N_{k+1} of M such that N_{k+1}/N_k is a minimal non-zero submodule of M/N_k. Since N_{k+1}/N_k is minimal in the set of non-zero submodules of M/N_k, then it must be irreducible. Thus in the chain $N_0 \subseteq \dots \subseteq N_k \subseteq N_{k+1}$, the quotient of two adjacent terms is irreducible. This process cannot be continued forever, since M is also noetherian. Thus we get a chain

$$0 = N_0 \subseteq \dots \subseteq N_r = M$$

of submodules of M such that N_i/N_{i-1} is irreducible for $i = 1, \dots, r$. Such a chain is called a *composition series* of M. It follows from the Artin-Schreier theorem [Cohn **84**, 9.2 Theorem 2] that any two composition series of M have the same number of submodules. The number r is called the *length* of M.

Since a holonomic module is noetherian and artinian, it must have a composition series. Hence its length is well-defined. It may be calculated using the argument in the proof of 2.2.

2.3 SCHOLIUM. *The length of a holonomic module cannot exceed its multiplicity.*

2.4 COROLLARY. *A holonomic A_n-modules of multiplicity 1 is irreducible.*

PROOF: Let M be a left holonomic module of multiplicity 1 and suppose that it has a non-zero submodule N. Then N and M/N are also holonomic by Proposition 1.1. Hence by Theorem 9.3.2, we have that $m(M) = m(N) + m(M/N)$. Since $m(M) = 1$ and $N \neq 0$, we have that $M/N = 0$. Thus $M = N$ and M must be irreducible.

It is surprisingly easy to calculate the number of generators of a module that is artinian and noetherian.

2.5 THEOREM. *Let R be a simple left noetherian ring and M a finitely generated left R-module. If M is artinian but R is not artinian (as a left R-module), then M is a cyclic module.*

PROOF: The statement that we shall prove is even sharper: if M is generated by $u_1, \ldots, u_r$ then there exist $a_2, \ldots, a_r \in R$ such that $u_1 + \sum_2^r a_i u_i$ generates M. It is enough to prove the case $r = 2$. The general case follows by induction.

Suppose that M is generated by two elements, say u and v. Since M is artinian and noetherian, it must have finite length. We shall prove, by induction on the length of M, that there exists $a \in R$ such that $M = A(u + av)$.

If M has length 1, then it is irreducible and there is nothing to prove. Suppose that the result holds for all R-modules of length $< r$, and that M has length r.

In particular Rv has finite length, so it must have a non-zero irreducible submodule. This must be cyclic, hence generated by an element of the form cv, for some $c \in R$. Since M/Rcv has smaller length than M, we may use the induction hypothesis to construct an element $\lambda \in R$ such that

$$M = R(u + \lambda v) + Rcv.$$

It is better to rephrase this by saying that one may assume that $M = Ru + Rv$, with Rv an irreducible module.

Let $\phi : R \to M$ be the homomorphism of modules defined by $\phi(x) = xu$. If ϕ is injective, then $R \cong \mathrm{im}\phi$ is a non-artinian submodule of M, which contradicts Proposition 2.1. Hence $\ker \phi \neq 0$. Let d be a non-zero element of $\ker \phi$.

Since R is a simple ring, one has that $RdR = R$, and so $RdR \cdot v \neq 0$. In particular, there exists $b \in R$, such that $dbv \neq 0$. We claim that M is generated by $u + bv$. First note that $d(u + bv) = dbv$ is a non-zero element in Rv. Since Rv is irreducible, v is a multiple of dbv. Hence v, and therefore u, are elements of $R(u + bv)$, which must then equal M.

2.6 COROLLARY. *Holonomic modules are cyclic.*

PROOF: We know from Theorem 2.2 that a holonomic A_n-module must be artinian. We have also seen that A_n is not artinian. The result follows from Theorem 2.5.

3. FURTHER EXAMPLES.

In this section we construct a family of holonomic A_n-modules which is very important in applications. We begin with a technical lemma. Note that in this lemma we do *not* assume that the module is finitely generated: this will be proved as part of the lemma.

3.1 LEMMA. *Let M be a left A_n-module with a filtration Γ with respect to the Bernstein filtration of A_n. Suppose that there exist constants c_1, c_2 such that for $j \gg 0$*

$$\dim_K \Gamma_j \leq c_1 j^n / n! + c_2 (j + 1)^{n-1}.$$

Then M is a holonomic A_n-module whose multiplicity cannot exceed c_1. In particular M is finitely generated.

PROOF: The proof breaks up naturally into two parts.

Part 1: Every finitely generated submodule of M is holonomic of multiplicity $\leq c_1$.

Let N be a finitely generated submodule of M. Thus N admits a good filtration, say Ω. By Proposition 8.3.2, there exists a positive integer r such that $\Omega_j \subseteq \Gamma_{j+r} \cap N$. In particular, $\dim_K \Omega_j \leq \dim_K \Gamma_{j+r}$. Let $\chi(t)$ be the Hilbert polynomial of N for the filtration Ω. Using the polynomial bound of the hypothesis, we conclude that, for very large j,

$$\chi(j) \leq c_1 (j + r)^n / n! + c_2 (j + r)^{n-1}.$$

In particular the degree of $\chi(t)$ cannot exceed n. Hence $d(N) \leq n$. From the Bernstein inequality we conclude that $d(N) = n$ equals the degree of $\chi(t)$. Going back to the polynomial inequality above, we obtain that $m(N) \leq c_1$.

Part 2: M is finitely generated.

Consider an ascending chain $N_1 \subseteq N_2 \subseteq \cdots \subseteq N_r$ of finitely generated submodules of M. Each of these is holonomic. By Theorem 9.3.2, $m(N_i) = m(N_{i-1}) + m(N_i/N_{i-1})$, for $i = 1, 2, \ldots, r$. Adding them up one has

$$\sum_{2}^{r} m(N_i/N_{i-1}) + m(N_1) = m(N_r) \geq c_1.$$

In particular, all ascending chains of finitely generated submodules of M have less than c_1 steps. Thus M must be finitely generated. We may now apply Part 1 to M itself, and conclude that it is holonomic of multiplicity $\leq c_1$.

Let $K(X) = K(x_1, \ldots, x_n)$ be the field of rational functions. We may extend the left action of A_n on the polynomial ring $K[X]$ to the field of rational functions. The x_i continue to act by multiplication. The action of ∂_i on a rational function f/g is defined by the rule for differentiation of quotients, namely

$$\partial_i(f/g) = (\partial_i(f)g - f\partial_i(g))/g^2.$$

A routine calculation shows that this action satisfies the required properties. Note that this module is not finitely generated.

Suppose that a polynomial p is chosen in $K[X]$. Let $K[X, p^{-1}]$ denote the set of rational functions of the form f/p^r, where f is a polynomial. Note that a partial derivative of f/p^r has denominator p^{2r}. Hence these rational functions are preserved by partial differentiation and by multiplication by a polynomial. In other words, $K[X, p^{-1}]$ is a left A_n-submodule of $K(X)$.

3.2 THEOREM. *The A_n-module $K[X, p^{-1}]$ is holonomic and its multiplicity is $\leq (\deg(p) + 1)^n$.*

PROOF: To save on notation, put $M = K[X, p^{-1}]$. Suppose that p has degree m. Set

$$\Gamma_k = \{f/p^k : \deg(f) \leq (m+1)k\}.$$

We first check, in detail, that Γ is a filtration for M.

Let f/p^k be an element of M and assume that f has degree s. Then $f/p^k = f \cdot p^s/p^{s+k}$. But fp^s has degree $s(m+1)$, which is $\leq (m+1)(s+k)$. Hence $f/p^k \in \Gamma_{s+k}$. It follows that M is the union of all Γ_k, for $k \geq 0$.

Next suppose that $f/p^k \in \Gamma_k$. Equivalently, f has degree $\leq (m+1)k$. Multiplication by x_i increases the degree of f by 1, thus $x_i(f/p^k) = x_if p/p^{k+1} \in \Gamma_{k+1}$. Differentiating f/p^k with respect to x_i we get

$$(p\partial_i(f) - kf\,\partial_i(p))/p^{k+1}.$$

The numerator has degree $\leq (m+1)k + (m-1)$, so that $\partial_i(f/p^k) \in \Gamma_{k+1}$. This may be summed up as

$$B_1.\Gamma_k \subseteq \Gamma_{k+1}.$$

Since $B_i = B_1^i$ we also have that $B_i\Gamma_k \subseteq \Gamma_{i+k}$.

Finally, the dimension of Γ_k cannot exceed the dimension of the vector space of polynomials of degree $\leq (m+1)k$. Hence each Γ_k is finite dimensional. This concludes the proof that Γ is a filtration of M, and shows that

$$\dim_K \Gamma_k \leq \binom{(m+1)k + n}{n}.$$

Since the two terms of highest degree in k of this binomial number are

$$(m+1)^n k^n/n! \text{ and } (m+1)^{n-1}(n+1)nk^{n-1}/2(n!)$$

it follows that

$$\dim_K \Gamma_k \leq (m+1)^n k^n/n! + (m+1)^{n-1}(n+1)n(k+1)^{n-1}/n!$$

for very large values of k. By Lemma 3.1, M must be a holonomic module of multiplicity $\leq (m+1)$, as required.

We may use the theory we have developed so far to construct the Bernstein polynomial of a differential operator with polynomial coefficients. As before, we start with a polynomial $p \in K[x_1, \ldots, x_n]$. Let s be a new variable. We are going to construct a holonomic module over the ring $A_n(K(s))$, where

$K(s)$ is the field of rational functions on s. The generator of this module will be denoted by p^s. This is a *formal symbol* on which ∂_j acts by

$$\partial_j \cdot p^s = sp^{-1}\partial p/\partial x_j \cdot p^s.$$

It follows from this formula that $A_n(K(s))p^s$ is an $A_n(K(s))$-submodule of $K(s)[X, p^{-1}]p^s$. Now define an automorphism t of $K(s)[X, p^{-1}]p^s$ by the formula $t(s^i p^s) = (s+1)^i p \cdot p^s$. Note that this is $A_n(K)$-linear, but *not* $A_n(K(s))$-linear, see Exercise 5.8.

3.3 THEOREM. *Let* $p \in K[X]$. *There exist a polynomial* $B(s) \in K[s]$ *and a differential operator* $D(s)$ *in the polynomial ring* $A_n(K)[s]$ *such that*

$$B(s)p^s = D(s)pp^s.$$

PROOF: The $A_n(K(s))$-module $K(s)[X, p^{-1}]p^s$ is holonomic. The proof is like that of Theorem 3.2 and is left to the reader; see Exercise 4.6. Since $A_n(K(s))p^s$ is a submodule of $K(s)[X, p^{-1}]p^s$, it must be holonomic. In particular $A_n(K(s))p^s$ has finite length by Scholium 2.3. Thus the descending sequence

$$A_n(K(s))p^s \supseteq A_n(K(s))p \cdot p^s \supseteq A_n(K(s))p^2 \cdot p^s \supseteq \cdots$$

must terminate. This means that there exists $k > 0$ such that

$$p^k \cdot p^s \in A_n(K(s))p^{k+1} \cdot p^s.$$

Applying t^{-k} to both sides of this equation, we get $p^s \in A_n(K(s))p \cdot p^s$. Clearing denominators, we conclude that there exists a polynomial $B(s) \in K[s]$ such that

$$B(s)p^s \in A_n(K)[s]p \cdot p^s$$

and the theorem is proved.

The polynomial $B(s)$ and the operator $D(s)$ are not uniquely determined by p. However, the set of all the possible polynomials $B(s)$ satisfying Theorem 3.3 is an ideal of $K[s]$, as one can easily verify. The monic generator of this ideal is thus unique; it is called the *Bernstein polynomial* of p and denoted by $b_p(s)$.

The calculation of the Bernstein polynomial for a given p can be a very complicated affair. Here is a famous example that can actually be done by hand. Let $p = x_1^2 + \cdots + x_n^2$. If we denote by D the differential operator $\partial_1^2 + \cdots + \partial_n^2$ then

$$D \cdot p^{s+1} = 4(s+1)(s+n/2)p^s.$$

Thus $b_p(s) = 4(s+1)(s+n/2)$. Note that in this case D is independent of s.

The polynomial $b_p(s)$ was introduced by Bernstein in [Bernstein **71**] as part of his solution of Gelfand's problem on the meromorphic extension of a certain analytic function. Bernstein's work is discussed in detail in [Krause and Lenagan **85**, Ch. 8] and requires a basic knowledge of distributions.

The Bernstein polynomial is also very important in singularity theory; see [Malgrange **76**]. Building on work of Malgrange, Kashiwara showed that the roots of the Bernstein polynomial are strictly negative rational numbers; see [Kashiwara **76**]. Kashiwara's approach works over any analytic manifold. For details see [Granger and Maisonobe **93**, Ch. VI] and [Björk **79**, Ch. 6]. The explicit calculation of the Bernstein polynomial and its roots has been the subject of many papers; see for example [Yano **78**] and [Cassou-Nogués **86**].

4. Exercises.

4.1 Show that $A_n/A_n\partial_n$ is a torsion A_n-module.
Hint: Every class in $A_n/A_n\partial_n$ contains a representative of the form $\sum_0^r d_i x_n^i$, where $d_i \in A_{n-1}$. Show that this element is annihilated by $\partial_n^{r+1} \in A_n/A_n\partial_n$.

4.2 Let I be a non-zero left ideal of A_n. Show that I is not an artinian A_n-module.

4.3 Show that $A_n/A_n\partial_n$ is not an artinian A_n-module.
Hint: The modules

$$(A_n x_n^k + A_n\partial_n)/A_n\partial_n$$

form an infinite descending chain in $A_n/A_n\partial_n$.

4.4 Let $p \in K[X]$ be a non-constant polynomial. Is the module $K[X, p^{-1}]$ irreducible?

4.5 For $r \leq n$, let $P = \sum_1^r \partial_i^2 - \sum_{r+1}^n \partial_i^2$. Find a left ideal J of A_n such that $A_n P \subseteq J$ and A_n / J is holonomic.

4.6 Let $p \in K[X]$. Show that $K(s)[X, p^{-1}]p^s$ is a holonomic $A_n(K(s))$-module.

Hint: Let m be the degree of p. The K-vector spaces, $\Gamma_k = \{q \cdot p^{-k} \cdot p^s :$ $\deg(q) \leq (m+1)k\}$ give rise to a filtration of $K(s)[X, p^{-1}]$. What are their dimensions?

4.7 Let p be a polynomial in $K[x_1, \ldots, x_n]$. Denote by $A_n[s] \cdot p^s$ the submodule of $K[s, X, p^{-1}]$ generated by p^s over the polynomial ring $A_n[s]$. This is also an A_n-module, but in this case *it is not clear whether it is finitely generated!* Show that if p belongs to the ideal of $K[X]$ generated by its partial derivatives, then $A_n[s] \cdot p^s$ is finitely generated as an A_n-module.
Hint: Let $D = \sum_1^n \frac{\partial p}{\partial x_i} \partial_i$. Then $D \cdot f = sf$.

4.8 Let p and $M = A_n[s]p^s$ be as in Exercise 4.7. Let $t : M \to M$ be the map defined by $t(D(s) \cdot p^s) = D(s+1)p \cdot p^s$.

(1) Show that t is an endomorphism of M as an A_n-module but *not* as an $A_n[s]$-module.
(2) Show that $[t, s] = t$.
(3) Use (2) to show that M/tM is an $A_n[s]$-module, even though t is not $A_n[s]$-linear.

It is true that M/tM is a holonomic module, but this is harder to prove; see [Granger and Maisonobe **93**, Ch. VI].

4.9 Let $\lambda_1, \ldots, \lambda_n$ be real numbers. Set $f(x_1, \ldots, x_n) = \exp(\lambda_1 x_1 + \cdots + \lambda_n x_n)$.

(1) Find generators for the left ideal $J = \{P \in A_n(\mathbb{R}) : P \cdot f = 0\}$.
(2) Show that $A_n(\mathbb{R})/J$ is a holonomic module over $A_n(\mathbb{R})$.

CHAPTER 11
CHARACTERISTIC VARIETIES

In this chapter we give a geometrical interpretation of the dimension of an A_n-module. In order to do this we will have to use a number of results of algebraic geometry and linear symplectic geometry. All the algebraic geometry that we need can be found in [Hartshorne 77, Ch. 1]. Throughout the chapter we assume that the base field is $\mathbb{C}$.

1. THE CHARACTERISTIC VARIETY.

Let A_n be the n-th complex Weyl algebra and let M be a finitely generated left A_n-module with a *good* filtration Γ. Then $gr^\Gamma M$ is a finitely generated module over the polynomial ring S_n. Let $ann(M, \Gamma)$ stand for the annihilator of $gr^\Gamma M$ in S_n. Then $ann(M, \Gamma)$ is an ideal of S_n. It depends not only on M, but also on the choice of the good filtration Γ; see Exercise 4.1. The radical of $ann(M, \Gamma)$ however is independent of the filtration.

1.1 LEMMA. *Let Ω be another good filtration of M. Then*

$$rad(ann(M, \Gamma)) = rad(ann(M, \Omega)).$$

PROOF: The ideals $ann(M, \Gamma)$ and $ann(M, \Omega)$ are homogeneous (see Exercise 8.4.6), hence so are their radicals (Exercise 4.2). Choose a homogeneous element f of degree s in $rad(ann(M, \Gamma))$. There exists $d \in B_s$, the component of degree $\leq s$ of the Bernstein filtration, such that $f = \sigma_s(d)$.

Since $f \in rad(ann(M, \Gamma))$, there exists $m \in \mathbb{N}$ such that $f^m \in ann(M, \Gamma)$. Hence, $d^m \Gamma_i \subsetneqq \Gamma_{ms+i-1}$, for every $i \geq 0$. Iterating q times we get that

(1.2) $$d^{mq} \Gamma_i \subseteq \Gamma_{i+msq-q}.$$

On the other hand, by Proposition 8.3.2 there exists $k \geq 0$ such that

$$\Gamma_{i-k} \subseteq \Omega_i \subseteq \Gamma_{i+k},$$

for all $i \geq 0$. Together with (1.2) for $q = 2k+1$, this leads to

$$d^{m(2k+1)} \Omega_i \subseteq d^{m(2k+1)} \Gamma_{i+k} \subseteq \Gamma_{i+ms(2k+1)-k-1} \subseteq \Omega_{i+ms(2k+1)-1}.$$

Thus $d^{m(2k+1)}\Omega_i \subseteq \Omega_{i+ms(2k+1)-1}$, for all $i \geq 0$. Hence $f^{m(2k+1)} \in ann(M, \Omega)$, and so $rad(ann(M, \Gamma)) \subseteq rad(ann(M, \Omega))$. The opposite inclusion follows by swapping Γ and Ω.

The ideal $\mathcal{I}(M) = rad(ann(M, \Gamma))$ is called the *characteristic ideal* of M. It follows from Lemma 1.1 that it is independent of the good filtration Γ used to calculate it. In other words, $\mathcal{I}(M)$ is an invariant of M. Thus, so is the affine variety

$$Ch(M) = \mathcal{Z}(\mathcal{I}(M)) \subseteq \mathbb{C}^{2n}$$

that it determines. This variety is called the *characteristic variety* of M. Since $\mathcal{I}(M)$ is a homogeneous ideal (see Exercise 4.2), the variety $Ch(M)$ is homogeneous. This means that if $p \in Ch(M)$ and $\lambda \in \mathbb{C}$, then $\lambda p \in Ch(M)$. Note that $Ch(M)$ is a subvariety of $\mathbb{C}^{2n}$, since S_n is a polynomial ring in $2n$ variables!

Here is a simple example. Let $d \in A_n$ be an element of degree r and put $M = A_n/A_n d$. If $\mathcal{B}''$ is the filtration of M induced by the Bernstein filtration of A_n, then

$$gr^{\mathcal{B}''} M = S_n/S_n \sigma_r(d)$$

as we saw in Ch. 7, §5. Therefore, $ann(M, \mathcal{B}'') = S_n \sigma_r(d)$ and so $Ch(M) = \mathcal{Z}(\sigma_r(d))$ is a hypersurface.

1.3 PROPOSITION. *Let M be a finitely generated left A_n-module and N a submodule of M. Then*

$$Ch(M) = Ch(N) \cup Ch(M/N).$$

PROOF: Let Γ be a good filtration of M. It induces good filtrations Γ' and Γ'' in N and M/N respectively. By Lemma 7.5.1, there exists an exact sequence of finitely generated graded S_n-modules,

$$0 \longrightarrow gr^{\Gamma'} N \longrightarrow gr^{\Gamma} M \longrightarrow gr^{\Gamma''} M/N \longrightarrow 0.$$

Clearly

$$ann(M, \Gamma) \subseteq ann(N, \Gamma') \cap ann(M/N, \Gamma'')$$

and so $Ch(N) \cup Ch(M/N) \subseteq Ch(M)$, by [Hartshorne **77**, Ch. 1, Proposition 1.2]. The opposite inclusion follows from

$$ann(N, \Gamma)ann(M/N, \Gamma'') \subseteq ann(M, \Gamma).$$

Before we proceed with the study of the characteristic variety, let us review some basic facts about the dimension of an affine variety. Let J be an ideal of $S_n = \mathbb{C}[y_1, \ldots, y_{2n}]$ and put $V = \mathcal{Z}(J)$. If p is a point of V then the *Zariski tangent space* of V at p is the linear subspace of $\mathbb{C}^{2n}$ defined by the equations

$$\sum_{1}^{2n} \frac{\partial F}{\partial y_i}(p) y_i = 0$$

for all $F \in J$. This space is denoted by $T_p(V)$; it is a complex vector subspace of $\mathbb{C}^{2n}$.

The *dimension* of V equals $\inf\{\dim_{\mathbb{C}} T_p(V) : p \in V\}$. Actually one need not look at every point of V. If p is a *non-singular point* of V, then $\dim(V) = \dim_{\mathbb{C}} T_p(V)$. This is equivalent to the definition in terms of heights of prime ideals, see [Hartshorne **77**, Ch. 1, Exercise 5.10].

We may also define $T_p(V)$ in terms of linear forms on $\mathbb{C}^{2n}$. Let $d_p f$ be the linear form defined on the vector $Y = (y_1, \ldots, y_{2n})$ by

$$d_p f(Y) = \sum_{1}^{2n} \frac{\partial f}{\partial y_i}(p) y_i.$$

Consider the map $d_p : S_n \to (\mathbb{C}^{2n})^*$ defined by $d_p(f) = d_p f$. Let $f, g \in S_n$ and $\lambda \in \mathbb{C}$. Then d_p satisfies

$$d_p(f + \lambda g) = d_p f + \lambda d_p g,$$
$$d_p(f g) = f(p)d_p g + g(p)d_p f.$$

We may rephrase the definition of $T_p(V)$ using d_p, as follows: $u \in T_p(V)$ if and only if $d_p F(u) = 0$, for every $F \in \mathcal{I}(V)$. This may be improved, by restricting the elements of $\mathcal{I}(V)$ to a finite set. Suppose that $F_1, \ldots, F_m$ generate $\mathcal{I}(V)$. It follows from the properties of d_p stated above that $u \in T_p(V)$ if and only if

$$d_p F_1(u) = \cdots = d_p F_m(u) = 0.$$

The characterization of $T_p(V)$ in terms of d_p will be used in the next section.

The following theorem is an immediate consequence of the fact that if N is a graded module over S_n then the degree of its Hilbert polynomial is $\dim \mathcal{Z}(ann_{S_n} N)$; see [Hartshorne **77**, Ch. 1 Theorem 7.5].

1.4 THEOREM. *Let M be a finitely generated left module over A_n. Then $\dim Ch(M) = d(M)$.*

It is now very easy to show that if $d \neq 0$ is an operator in A_n, then $A_n/A_n d$ has dimension $2n-1$. Suppose that d has degree $r > 0$. As we have seen, the characteristic variety of $A_n/A_n d$ has equation $\sigma_r(d) = 0$. This is a hypersurface of $\mathbb{C}^{2n}$; hence its dimension is $2n-1$. Thus $d(A_n/A_n d) = 2n-1$, by Theorem 1.4. Compare this example with Exercise 9.5.2.

We may also use the characteristic variety to achieve a geometrical interpretation of Bernstein's inequality. This depends on results of symplectic geometry that we summarize in the next section.

2. SYMPLECTIC GEOMETRY.

A symplectic structure on $\mathbb{C}^{2n}$ is determined by a *non-degenerate skew-symmetric form*. The *standard* symplectic structure on $\mathbb{C}^{2n}$ is expressed by means of the matrix

$$\Omega = \begin{pmatrix} 0 & -I_n \\ I_n & 0 \end{pmatrix}$$

where I_n is the $n \times n$ identity matrix. Given vectors $u, v \in \mathbb{C}^{2n}$, the symplectic form is

$$\omega(u, v) = u\Omega v^t.$$

The matrix of any non-degenerate skew-symmetric form can be written in the form above for a suitable choice of coordinates [Cohn **84**, §8.6, Theorem 1].

If U is a subspace of $\mathbb{C}^{2n}$, then its *skew-orthogonal complement* is

$$U^\perp = \{v \in \mathbb{C}^{2n} : \omega(u, v) = 0 \text{ for all } u \in U\}.$$

Note that it is *not* true that $\mathbb{C}^{2n}$ is the direct sum of a subspace and its skew-orthogonal complement. For example, since $\omega(u, u) = 0$ for every $u \in \mathbb{C}^{2n}$, then every one-dimensional space is contained in its complement. A subspace that is contained in its skew-symmetric complement is called *isotropic*. However, since ω is non-degenerate, it is always true that U and $U^\perp$ have complementary dimensions. The proof is short enough to be included here. First, some definitions. For $w \in \mathbb{C}^{2n}$, define a linear form $\phi_w : \mathbb{C}^{2n} \to \mathbb{C}$ by $\phi_w(v) = \omega(w, v)$. Let $\Phi : \mathbb{C}^{2n} \to (\mathbb{C}^{2n})^*$ be the linear map which associates ϕ_w to $w \in \mathbb{C}^{2n}$.

2.1 PROPOSITION. *Let ω be a non-degenerate skew-symmetric form in $\mathbb{C}^{2n}$ and U a subspace of $\mathbb{C}^{2n}$. Then:*

(1) *Φ is an isomorphism.*

(2) *If a linear form ϕ_w restricts to zero on U, then $w \in U^\perp$.*

(3) $\dim U + \dim U^\perp = 2n$.

PROOF: Note that the kernel of Φ is zero, because ω is non-degenerate. Hence,

$$\dim \Phi(\mathbb{C}^{2n}) = 2n - \dim \ker(\Phi) = 2n.$$

Since $\dim(\mathbb{C}^{2n})^* = 2n$, it follows that Φ is surjective. Thus Φ is an isomorphism of vector spaces, and (1) is proved.

Now let $\Phi|_U : \mathbb{C}^{2n} \to U^*$ be the map defined by $\Phi|U(w) = \phi_w|_U$, the restriction of ϕ_w to U. Since every linear form on U extends to a linear form on $\mathbb{C}^{2n}$, we have that $\Phi|_U$ is surjective by (1). Hence the image of $\Phi|_U$ is U^*. Now (2) is clear, and it implies that the kernel of $\Phi|_U$ is $U^\perp$. Hence,

$$2n = \dim \ker \Phi|_U + \dim \Phi|_U(\mathbb{C}^{2n}) = \dim U^\perp + \dim U^*,$$

which implies (3).

The subspaces that are important, from our point of view, are the ones that contain their skew-complement. They are called *co-isotropic* or *involutive*. A hyperplane is always an involutive subspace. Indeed, if H is a hyperplane of $\mathbb{C}^{2n}$ that is not involutive, then there exists $v \notin H$ such that $\omega(v, u) = 0$, for all $u \in H$. Since $\mathbb{C}^{2n} = H + \mathbb{C}v$, this contradicts the non-degeneracy of ω.

If V is an affine variety of $\mathbb{C}^{2n}$, then we will say that it is *involutive* if the tangent space $T_p(V) \subseteq \mathbb{C}^{2n}$ is involutive at every non-singular point $p \in V$.

In particular, a hypersurface will always be involutive. The dimension of an involutive variety satisfies an inequality similar to Bernstein's.

2.2 PROPOSITION. *The dimension of an involutive variety of $\mathbb{C}^{2n}$ is greater than or equal to n.*

PROOF: Let V be an involutive variety of $\mathbb{C}^{2n}$ and p a non-singular point of V. By definition, we have that $T_p(V)^\perp \subseteq T_p(V)$. Hence $\dim T_p(V)^\perp \leq \dim T_p(V)$. Thus, by Proposition 2.1,

$$2n = \dim T_p(V)^\perp + \dim T_p(V) \leq 2 \dim T_p(V),$$

and so $\dim T_p(V) \geq n$. Therefore, $\dim(V) \geq n$.

The involutive varieties of dimension n are called *lagrangian*. Note that if V is lagrangian in $\mathbb{C}^{2n}$, then $T_p(V)^\perp \subset T_p(V)$ and both subspaces have dimension n. Hence $T_p(V) = T_p(V)^\perp$. In particular, V is also isotropic. Thus lagrangian varieties can also be characterized as varieties which are involutive (co-isotropic) *and* isotropic.

Talk of an affine variety, and one immediately thinks of its defining ideal. How can one decide whether a variety is involutive by looking at its ideal? The answer lies with the Poisson bracket of S_n.

Let I denote the inverse of the map Φ defined above. The *Poisson bracket* of $f, g \in S_n$ is

$$\{f, g\}(p) = \omega(I\,d_p f, I\,d_p g) = d_p f(I\,d_p g).$$

An explicit calculation using coordinates shows that

$$\{f, g\}(p) = \sum_1^n \{\frac{\partial f}{\partial y_{n+i}}(p) \cdot \frac{\partial g}{\partial y_i}(p) - \frac{\partial g}{\partial y_{n+i}}(p) \cdot \frac{\partial f}{\partial y_i}(p)\}.$$

From this formula it is easy to see that $\{f, g\}$ is a polynomial in S_n and that the map $\{f, \cdot\} : S_n \to S_n$ is a derivation of S_n. An ideal J of S_n is *closed* for the Poisson bracket if $\{f, g\} \in J$ whenever $f, g \in J$. We want to show that a variety is involutive if its ideal is closed for the Poisson bracket. The proof will make use of a technical fact about the tangent space that we isolate in a lemma.

2.3 LEMMA. *Let V be an affine variety of $\mathbb{C}^{2n}$ and p a point of V. If θ is a form on $\mathbb{C}^{2n}$ whose restriction to $T_p(V)$ is zero, then there exists $f \in \mathcal{I}(V)$ such that $\theta = d_p f$.*

PROOF: Let $F_1, \ldots, F_m$ be the generators of $\mathcal{I}(V)$. Set $Y = (y_1, \ldots, y_{2n})$. Then $T_p(V)$ is the solution set of the equations

$$d_p F_1(Y) = \cdots = d_p F_m(Y) = 0$$

in $\mathbb{C}^{2n}$. But $\theta(Y) = 0$ is satisfied by the vectors of $T_p(V)$, by hypothesis. Thus

$$\theta = \sum_1^m a_i d_p F_i$$

where $a_i \in \mathbb{C}$. Let $f = \sum_1^m a_i F_i \in \mathcal{I}(V)$. A straightforward calculation shows that $\theta = d_p f$.

2.4 PROPOSITION. *An affine variety V of $\mathbb{C}^{2n}$ is involutive if and only if its ideal $\mathcal{I}(V)$ is closed for the Poisson bracket.*

PROOF: Suppose that V is involutive. Let p be a non-singular point of V. If $f \in \mathcal{I}(V)$, then $d_p f$ restricted to $T_p(V)$ is zero. By Proposition 2.1(2), $I d_p f \in T_p(V)^\perp$. Since V is involutive, then $I d_p f \in T_p(V)$. Now if $g \in \mathcal{I}(V)$, then $d_p g$ is zero on $T_p(V)$, hence

$$\{f, g\}(p) = -d_p g(I d_p f) = 0.$$

Since this identity holds for all non-singular points p of V, we conclude that the polynomial $\{f, g\}$ is in $\mathcal{I}(V)$, which is then closed for the Poisson bracket.

Conversely, assume that $\mathcal{I}(V)$ is closed for the Poisson bracket. Choose $w \in T_p(V)^\perp$. By Lemma 2.3, there exists $g \in \mathcal{I}(V)$ such that $\phi_w = d_p g$. But $d_p g(u) = \omega(I d_p g, u)$. Since ω is non-degenerate, $w = I d_p g$. If $f \in \mathcal{I}(V)$ then

$$0 = \{f, g\}(p) = d_p f(I d_p g) = d_p f(w).$$

Hence $w \in T_p(V)$. Thus $T_p(V)^\perp \subseteq T_p(V)$.

We have seen that hypersurfaces are involutive, since their tangent spaces are hyperplanes. This is very easy to prove using Proposition 2.4, since the ideal of a hypersurface is principal, and the Poisson bracket of a polynomial with itself is always zero.

The relation between characteristic varieties and symplectic geometry is the subject of the next theorem. The first proofs of this result were analytical. There is now a purely algebraic proof due to O. Gabber [Gabber **81**]. We shall not include the proof here as it would take us too far from our intended course.

2.5 THEOREM. *Let M be a finitely generated left A_n-module. Then $Ch(M)$ is involutive with respect to the standard symplectic structure of $\mathbb{C}^{2n}$. Equivalently, $\mathcal{I}(M)$ is closed for the Poisson bracket.*

It is not obvious at first sight why this theorem should be difficult to prove. In fact, it is easy to prove that if Γ is a good filtration of M, then $ann(M, \Gamma)$ is closed for the Poisson bracket. The theorem does *not* follow easily from this fact because it is *not* true that the radical of a closed ideal is closed! Here is an example. Let J be the ideal of S_1 generated by y_1^2, y_2^2 and $y_1 y_2$. It is closed for the Poisson bracket, but its radical contains y_1 and y_2, and $\{y_1, y_2\} = 1$. Thus $rad(J)$ is not closed for the Poisson bracket. Theorem 2.5 states the very subtle fact that the radical of the annihilator of the graded module of an A_n-module *is* closed, even though radicals of closed ideals are *not* in general closed.

Putting together Theorem 2.5 and Proposition 2.2 we get Bernstein's inequality.

2.6 COROLLARY. *Let M be a finitely generated left A_n-module. Then*

$$d(M) = \dim Ch(M) \geq n.$$

It also follows from Theorem 2.5 that a holonomic module has a homogeneous lagrangian characteristic variety. A very spectacular use of the involutivity of the characteristic variety is discussed in the next section.

3. NON-HOLONOMIC IRREDUCIBLE MODULES

It is a very curious fact that, until 1985, it was widely believed that every irreducible A_n-module had to be holonomic. There was no good reason for this, except a lack of examples. The truth came to light in [Stafford 85]. Stafford showed that if $n > 2$ and $\lambda_2, \ldots, \lambda_n \in \mathbb{C}$ are algebraically independent over $\mathbb{Q}$, then the operator

$$s = \partial_1 + (\sum_2^n \lambda_i x_1 x_i \partial_i + x_i) + \sum_2^n (x_i - \partial_i)$$

generates a maximal left ideal of A_n. Let $M = A_n/A_n s$. Then M is irreducible and $d(M) = 2n - 1 > n$, since $n > 2$. The fact that $A_n s$ is maximal is proved by a long

calculation, which can also be found in [Krause and Lenagan 85, Theorem 8.7].

In 1988, J. Bernstein and V. Lunts found a different and more geometric way of constructing irreducible modules of dimension $2n - 1$ over A_n. As we have seen in §§1 and 2, the characteristic variety of an A_n-module is always homogeneous and involutive. We will say that a homogeneous involutive variety of $\mathbb{C}^{2n}$ is *minimal* if it does not contain a proper involutive homogeneous subvariety. For example, an irreducible lagrangian variety must be minimal, since varieties of dimension less than n cannot be involutive. The work of Bernstein and Lunts depends on the following result.

3.1 PROPOSITION. *Let d be an operator of degree k in A_n and suppose that:*

(1) *the symbol $\sigma_k(d)$ is irreducible in S_n;*
(2) *the hypersurface $\mathcal{Z}(\sigma_k(d))$ is minimal.*

Then the left ideal $A_n d$ is maximal. In particular, the quotient $A_n/A_n d$ is an irreducible module of dimension $2n - 1$ over A_n.

PROOF: Suppose that J is a left ideal of A_n such that $A_n d \subset J$. Since these are submodules of A_n we may consider their graded modules with respect to the filtrations induced by $\mathcal{B}$; we get

$$S_n \sigma_k(d) \subset gr(J) \subseteq S_n.$$

The corresponding varieties are

$$\emptyset \subseteq \mathcal{Z}(gr(J)) \subset \mathcal{Z}(\sigma_k(d)).$$

Note that the last inclusion is proper, because $\sigma_k(d)$ is irreducible. But $gr(A_n/J) \cong S_n/gr(J)$, and so $ann(A_n/J, \mathcal{B}) = gr(J)$. Therefore, $\mathcal{Z}(gr(J)) = Ch(A_n/J)$ is involutive by Theorem 2.5. Since $\mathcal{Z}(\sigma_k(d))$ is minimal, then $\mathcal{Z}(gr(J)) = \emptyset$. Hence $gr(J) = S_n$ and so $J = A_n$. Thus A_nd is a maximal left ideal of A_n.

To put this proposition to good use we have only to construct minimal hypersurfaces in $\mathbb{C}^{2n}$. That these hypersurfaces exist is the heart of the work of Bernstein and Lunts. In fact they show that most hypersurfaces in $\mathbb{C}^{2n}$ are minimal. To make this into a precise statement we need a definition.

Let $S_n(k)$ be the homogeneous component of degree k of S_n. This is a finite dimensional complex vector space; so it makes sense to talk about hypersurfaces in $S_n(k)$. We say that a property **P** holds *generically* in $S_n(k)$ if the set

$$\{f \in S_n(k) : \textbf{P} \text{ does not hold for } f\}$$

is contained in the union of countably many hypersurfaces of $S_n(k)$.

3.2 THEOREM. *The property '$\mathcal{Z}(f)$ is minimal' holds generically in $S_n(k)$, whenever $k \geq 4$ and $n \geq 2$.*

This result was proved for $n = 2$ in [Bernstein and Lunts **88**] and later generalized to $n \geq 2$ in [Lunts **89**]. The proof of Theorem 3.2 uses a lot of algebraic geometry and also some results on differential equations, one of which goes back to Poincaré's thesis! The proficient reader will find more details in the original papers.

The gist of the work of Bernstein and Lunts is that 'most' irreducible A_n-modules are *not* holonomic: almost the exact opposite of what was believed before 1985.

4. EXERCISES

4.1 Let $M = A_1/A_1x$. Let Γ be the filtration of M induced by $\mathcal{B}$ and Ω the filtration defined by $\Omega_k = B_k \cdot (\partial + A_1x)$. Show that $ann(M, \Gamma) = S_1y_1$ and $ann(M, \Omega) = S_1y_1^2 + S_1y_1y_2$. Compute their radicals.

4.2 Show that if J is a homogeneous ideal of a graded algebra R then $rad(J)$ is also homogeneous.

4.3 Let J be a left ideal of A_n and put $M = A_n/J$. Show that $Ch(M) = \mathcal{Z}(gr(J))$.

4.4 Is the union of two involutive varieties involutive? What about their intersection?

4.5 Show that if V is an involutive homogeneous variety of $\mathbb{C}^{2n}$ then its irreducible components are also homogeneous and involutive.

4.6 Let **s** be the operator of A_n defined in §3. Let $M = A_n/A_n\mathbf{s}$.

   (1) Compute $Ch(M)$.

   (2) Is it an irreducible variety of $\mathbb{C}^{2n}$?

   (3) Is it a minimal hypersurface?

4.7 Let J be a left ideal of A_n. Show that if $gr(J)$ is a prime ideal of S_n and $\mathcal{Z}(gr(J))$ is minimal, then J is a maximal ideal of A_n.

4.8 When is a hypersurface of $\mathbb{C}^{2n}$ lagrangian? Give an example of a lagrangian variety of $\mathbb{C}^{2n}$, when $n \geq 2$.

4.9 A holonomic left A_n-module M is *regular* if there exists a filtration Γ for M such that $ann_{S_n}gr^\Gamma M$ is a radical ideal of S_n. Let N be a submodule of a regular holonomic module M. Show that N and M/N are also regular.

4.10 Show that if M is a regular holonomic module whose characteristic variety is irreducible, then M is an irreducible module. Why is the *regular* hypothesis required?

4.11 Let J be an ideal of S_n. Show that J^2 is always closed for the Poisson bracket.

CHAPTER 12

TENSOR PRODUCTS

All the operations on A_n-modules to be defined in the next chapters make use of the tensor product, which we are about to study. The construction of the tensor product presented in §2 usually seems artificial on a first encounter. Fear not; it is the neat *universal property* of §3 that is most often used in the applications. The final two sections before the exercises contain a number of results that will be required later.

1. BIMODULES.

To discuss tensor products in sufficient generality it is necessary to introduce bimodules. Let R and S be rings and let M be an abelian group. To qualify as an R-S-*bimodule*, the group M must be a left R-module and a right S-module, and the R-action and the S-action must be compatible in the sense that if $r \in R$, $s \in S$ and $u \in M$, then

$$r(us) = (ru)s.$$

If $S = R$ then we simply say that M is an R-bimodule.

A few examples and counter-examples will make the definition clear. If R is a ring then R^k is an R-bimodule. More generally, if R is a subring of S, then S^k is an S-R bimodule. In particular this applies to Weyl algebras; for if $m < n$, then A_m is a subring of A_n. Thus A_n^k is an A_n-A_m-bimodule.

On the other hand, $K[x]$ is a left A_1-module and a right $K[x]$-module, but it is not an A_1-$K[x]$-bimodule because the two actions are not compatible. Indeed, suppose that $f \in K[x]$ is acted on the left by ∂ and on the right by x. Then

$$\partial(fx) = x\partial(f) + f$$

but $(\partial \cdot f)x = x\partial(f)$. More generally, $K[X]$ is *not* an A_n-$K[X]$-bimodule.

Let M be an R-S-bimodule. A subgroup N of M is a *sub-bimodule* of M if it is stable under both the R-action and the S-action on M. In this case, the quotient group M/N has a natural structure of R-S-bimodule.

The following example will often come up in applications. Consider A_n as a subring of A_{n+1} in the usual way. The left ideal $A_{n+1}x_{n+1}$ is a sub-bimodule of the A_{n+1}-A_n-bimodule A_{n+1}. Thus the quotient

$$A_{n+1}/A_{n+1}x_{n+1}$$

is an A_{n+1}-A_n-bimodule.

We may similarly define a homomorphism of R-S-bimodules as a homomorphism of the underlying abelian groups which preserves both the left and right module structures.

2. Tensor products.

Let R, S and T be rings. Let M be an R-S-bimodule and let N be an S-T-bimodule. We will define the tensor product of M and N over S, denoted by $M \otimes_S N$.

First consider the set $M \times N$ of all pairs (u, v), with $u \in M$, $v \in N$. Let $\mathcal{A}$ be the free abelian group whose basis is formed by the elements of $M \times N$. The elements of $\mathcal{A}$ are formal (finite) sums of the form

(2.1) $$\sum_i a_i(u_i, v_i)$$

with $a_i \in \mathbb{Z}$, $u_i \in M$, $v_i \in N$. Note that in this sum the pairs are mere symbols: the sum is *not* an element of the direct sum $M \oplus N$. In fact, if we assume that different indices correspond to different pairs in (2.1), then the sum is zero if and only if each $a_i = 0$. If $r \in R$ and $t \in T$, put

$$r(u, v) = (ru, v),$$
$$(u, v)t = (u, vt),$$

where $u \in M$ and $v \in N$. These are well-defined actions that make an R-T-bimodule of $\mathcal{A}$. Now consider the subgroup $\mathcal{B}$ of $\mathcal{A}$ generated by elements of the following types:

$$(u + u', v) - (u, v) - (u', v),$$
$$(u, v + v') - (u, v) - (u, v'),$$

$$(us, v) - (u, sv),$$

where $r \in R$, $s \in S$, $t \in T$, $u, u' \in M$, $v, v' \in N$. One immediately checks that $\mathcal{B}$ is a sub-bimodule of $\mathcal{A}$.

The tensor product $M \otimes_S N$ is the quotient $\mathcal{A}/\mathcal{B}$. The image of (u, v) in $M \otimes_S N$ is denoted by $u \otimes v$. Since $\mathcal{A}$ has the pairs (u, v) as basis, the elements of the form $u \otimes v$ generate $M \otimes_S N$. However, it is *not* true that every element of the tensor product is of the form $u \otimes v$.

The elements $u \otimes v$ satisfy relations which are determined by the generators of $\mathcal{B}$, as follows

$$(u + u') \otimes v = u \otimes v + u' \otimes v,$$
$$u \otimes (v + v') = u \otimes v + u \otimes v',$$

which are a sort of distributivity; and also

$$us \otimes v = u \otimes sv,$$

which means that the tensor product is *balanced*; finally

$$r(u \otimes v) = ru \otimes v,$$
$$(u \otimes v)t = u \otimes vt,$$

are, respectively, the left R-action and right T-action on $M \otimes_S N$.

Note that if R is a K-algebra then a left R-module may be seen as an R-K-bimodule. This trivial observation must be kept in mind in the applications. For example, if M is a left R-module and N a right R-module, then $N \otimes_R M$ is a K-vector space. We may also calculate $M \otimes_K N$, which is an R-bimodule. On the other hand, if R is commutative then an R-module is automatically an R-bimodule. In particular if M and N are R-modules and R is commutative, then $M \otimes_R N$ is an R-module.

Let us end this section by calculating a tensor product using the definition above. This is a simple example which, at the same time, carries a warning: the tensor product can behave in very unexpected ways. Consider the $\mathbb{Z}$-bimodules $\mathbb{Z}_p$ and $\mathbb{Z}_q$, and assume that p and q are co-prime. We wish to calculate $\mathbb{Z}_p \otimes_{\mathbb{Z}} \mathbb{Z}_q$. We know that it is generated by elements of the form

$m \otimes n$, where $m \in \mathbb{Z}_p$, $n \in \mathbb{Z}_q$. But since p and q are co-prime, there exist integers a, b such that $ap + bq = 1$. Thus

$$m \otimes n = (ap + bq)m \otimes n = bq \cdot m \otimes n$$

because $pm = 0$. Since the product is balanced, we deduce from the above that $m \otimes n = bm \otimes qn$. But $qn = 0$, and so we end up with $m \otimes n = 0$. Therefore, $\mathbb{Z}_p \otimes_\mathbb{Z} \mathbb{Z}_q = 0$.

3. THE UNIVERSAL PROPERTY

The first property of the tensor product that we shall study is also the key to proving all the other ones. To make its statement more succinct it is convenient to introduce some terminology. For the rest of this section let R, S and T be rings; let M be an R-S-bimodule and N an S-T-bimodule. Suppose that L is an R-T-bimodule and that $\phi : M \times N \longrightarrow L$ is a map. We say that ϕ is *bilinear* if

$$\phi(ru + u', v) = r\phi(u, v) + \phi(u', v),$$
$$\phi(u, vt + v') = \phi(u, v)t + \phi(u, v'),$$

for all $u, u' \in M$, $v, v' \in N$, $r \in R$, $t \in T$. The map ϕ is said to be *balanced*, if

$$\phi(us, v) = \phi(u, sv)$$

for all $s \in S$, $u \in M$, $v \in N$. The canonical example of a bilinear and balanced map is the projection

$$\pi : M \times N \longrightarrow M \otimes_S N$$

defined by $\pi(u, v) = u \otimes v$.

Let U be an R-T-bimodule, and suppose that there is a bilinear and balanced map $\eta : M \times N \longrightarrow U$. We say that η is a *universal* bilinear balanced map if given any R-T-bimodule L and a bilinear and balanced map $\phi : M \times N \longrightarrow L$, there exists a unique R-T-bimodule homomorphism $\overline{\phi} : U \longrightarrow L$ such that $\phi = \overline{\phi} \cdot \eta$.

3.1 THEOREM. *The map* $\pi : M \times N \longrightarrow M \otimes_S N$ *is a universal bilinear balanced map.*

PROOF: First of all, we may extend ϕ to get a map $\psi : \mathcal{A} \longrightarrow L$. In detail,

$$\psi(\sum_i a_i(u_i, v_i)) = \sum_i a_i\phi(u_i, v_i).$$

Since ϕ is bilinear and balanced, it follows that ψ is zero on every element of the submodule $\mathcal{B}$. Thus ψ induces an R-T-bimodule map

$$M \otimes_S N = \mathcal{A}/\mathcal{B} \longrightarrow L.$$

This is the map that we call $\overline{\phi}$. Since the canonical projection of $\mathcal{A}$ onto $\mathcal{A}/\mathcal{B}$ is induced by π, the equation $\phi = \overline{\phi} \cdot \pi$ follows immediately. We must now show that $\overline{\phi}$ is unique. Note that any bilinear and balanced map $\theta : M \otimes_S N \longrightarrow L$ for which $\phi = \theta \cdot \pi$ satisfies $\theta(u \otimes v) = \phi(u, v)$, for any $u \in M$, $v \in N$. But this equation is enough to characterize $\overline{\phi}$; hence $\theta = \overline{\phi}$.

It follows from this theorem that the universal property uniquely characterizes the tensor product.

3.2 COROLLARY. *Let* $\eta : M \times N \longrightarrow U$ *be a universal bilinear balanced map. Then* $U \cong M \otimes_S N$ *as* R-T-bimodules.

PROOF: Since π is a universal bilinear balanced map, there exists $\overline{\eta} : M \otimes_S N \longrightarrow U$ such that $\eta = \overline{\eta} \cdot \pi$. On the other hand, since η is also universal, there exists $\overline{\pi} : U \longrightarrow M \otimes_S N$ such that $\pi = \overline{\pi} \cdot \eta$. Thus $\eta = (\overline{\eta} \cdot \overline{\pi}) \cdot \eta$. Using again the universality of η, this time with respect to itself, and its uniqueness, we get that $\overline{\eta} \cdot \overline{\pi} = \iota$, where $\iota : U \longrightarrow U$ is the identity map. Similarly, $\overline{\pi} \cdot \overline{\eta}$ is the identity on $M \otimes_S N$.

We must now consider what happens with maps when we take tensor products. Let M' be an R-S-bimodule and N' be an S-T-bimodule. Let $\phi : M \longrightarrow M'$ and $\psi : N \longrightarrow N'$ be two bimodule homomorphisms. Consider the map $\theta : M \times N \longrightarrow M' \otimes_S N'$ defined by $\theta(u, v) = \phi(u) \otimes \psi(v)$. It follows from the properties of the tensor product that this map is bilinear and balanced. Hence, by the universal property of bilinear balanced maps, there exists a bimodule homomorphism:

$$\overline{\theta} : M \otimes_S N \longrightarrow M' \otimes_S N'.$$

This map is usually denoted by $\phi \otimes \psi$. Note that

$$(\phi \otimes \psi)(u \otimes v) = \phi(u) \otimes \psi(v).$$

4. BASIC PROPERTIES.

We now use the universal property to prove some of the basic properties of the tensor product of modules. We illustrate the method with some examples, and leave the others as exercises for the reader.

4.1 PROPOSITION. *Let R, S be rings, and M be an R-S-bimodule. Then*

$$R \otimes_R M \cong M \cong M \otimes_S S.$$

PROOF: Let θ be the map from $R \times M$ to M defined by $\theta(r, u) = ru$. It follows from the definition of an R-module that θ is bilinear and balanced. By the universal property of the tensor product there exists a bimodule homomorphism $\bar{\theta}$ from $R \otimes_R M$ to M. This map satisfies the equation $\bar{\theta}(r \otimes u) = ru$. It is clear that $\bar{\theta}$ is onto, we show that it is injective. Since we are tensoring up over R, we have that $r \otimes u = 1 \otimes ru$. Thus every element of $R \otimes_R M$ can be written in the form $1 \otimes u$, for some $u \in M$. But $\bar{\theta}(1 \otimes u) = u$, hence $\bar{\theta}$ must be injective. The proof of the other isomorphism is analogous and is left to the reader.

The property that we now prove is the distributivity of the tensor product with respect to the direct sum. It will be used often in later chapters.

4.2 THEOREM. *Let R, S, T be rings, and $\mathcal{I}$ be an index set. Suppose that M is an R-S-bimodule and that N_i are S-T-bimodules, for every $i \in \mathcal{I}$. Then*

$$M \otimes_S \bigoplus_i N_i \cong \bigoplus_i (M \otimes_S N_i).$$

PROOF: The elements of $\bigoplus_i N_i$ will be denoted by $\bigoplus_i v_i$, where $v_i \in N_i$ for every $i \in \mathcal{I}$. Consider the map

$$\theta : M \times \bigoplus_i N_i \longrightarrow \bigoplus_i (M \otimes_S N_i)$$

defined by $\theta(u, \bigoplus_i v_i) = \bigoplus_i (u \otimes v_i)$. It is a bilinear and balanced map. By the universal property of the tensor product, there exists a bimodule map

$$\bar{\theta} : M \otimes_S \bigoplus_i N_i \longrightarrow \bigoplus_i (M \otimes_S N_i)$$

which satisfies $\bar{\theta}(u \otimes (\bigoplus_i v_i)) = \bigoplus_i (u \otimes v_i)$.

On the other hand, let $\mu_j : N_j \to \bigoplus_i N_i$ be the natural embedding. Thus we have maps

$$Id \otimes \mu_j : M \otimes N_j \longrightarrow M \otimes_S \bigoplus_i N_i.$$

By the universal property of direct sums [Cohn **84**, p.311], there exists a unique map

$$\mu : \bigoplus_i (M \otimes N_i) \longrightarrow M \otimes (\bigoplus_i N_i)$$

such that $\mu(\bigoplus_i (u \otimes v_i)) = u \otimes (\bigoplus_i v_i)$. One now checks that μ and $\bar{\theta}$ are inverse to each other by a simple calculation on generators; hence both maps are isomorphisms.

A similar result holds when the direct sum is the first entry of the tensor product. Thus the following corollary is an immediate consequence of Theorem 4.2.

4.3 COROLLARY. *Let R be a commutative ring and $\mathcal{I}_1$, $\mathcal{I}_2$ be index sets. Let F_i be free R-modules with bases $\{v_\alpha^i : \alpha \in \mathcal{I}_i\}$, for $i = 1, 2$. Then $F_1 \otimes_R F_2$ is free with basis $\{v_\alpha^1 \otimes v_\beta^2 : \alpha \in \mathcal{I}_1, \beta \in \mathcal{I}_2\}$. Furthermore, if F_1 and F_2 have finite bases, then*

$$rank(F_1 \otimes_R F_2) = rank(F_1)rank(F_2).$$

Let us now state a few more properties of the tensor product: the proofs follow the general pattern of 4.1 and 4.2.

4.4 PROPOSITION. *Let R_i be rings and M_i be R_i-R_{i+1}-bimodules, for $i = 1, 2, 3$. Then*

$$M_1 \otimes_{R_2} (M_2 \otimes_{R_3} M_3) \cong (M_1 \otimes_{R_2} M_2) \otimes_{R_3} M_3.$$

4.5 PROPOSITION. *Let R be a commutative ring and let M and N be R-modules. Then*

$$M \otimes_R N \cong N \otimes_R M.$$

We must also consider the effect of the tensor product on exact sequences. The first result holds in great generality.

4.6 THEOREM. *Let R, S and T be rings and let*

$$M' \xrightarrow{\psi} M \xrightarrow{\phi} M'' \to 0$$

be an exact sequence of S-T-bimodules. If B is an R-S-bimodule, then the sequence

$$B \otimes_S M' \xrightarrow{1 \otimes \psi} B \otimes_S M \xrightarrow{1 \otimes \phi} B \otimes_S M'' \to 0$$

is exact.

PROOF: Since ϕ is surjective, the elements of $B \otimes_S M''$ can be written in the form

$$\sum_1^k b_i \otimes \phi(u_i)$$

where $b_i \in B$ and $u_i \in M$. Since this is equal to

$$(1 \otimes \phi) \left(\sum_1^k b_i \otimes u_i \right)$$

we conclude that $1 \otimes \phi$ is surjective. On the other hand,

$$(1 \otimes \phi) \cdot (1 \otimes \psi) = (1 \otimes \phi \cdot \psi)$$

equals zero, since $\ker(\phi) = \operatorname{im}(\psi)$. Hence

$$\operatorname{im}(1 \otimes \psi) \subseteq \ker(1 \otimes \phi)$$

and the proof will be complete if we show that the opposite inclusion holds.
Let $N = \operatorname{im}(1 \otimes \psi)$. Then ϕ induces a map,

$$\theta : (B \otimes_S M)/N \to B \otimes_S M''$$

defined by $\theta(b \otimes u + N) = b \otimes \phi(u)$. Note that θ is well-defined because $N \subseteq \ker(1 \otimes \phi)$. Now let π be the projection

$$\pi : B \otimes_S M \to (B \otimes_S M)/N.$$

One easily checks that $\theta \cdot \pi = 1 \otimes \phi$. Suppose we have shown that θ is an isomorphism, then

$$\ker(1 \otimes \phi) = \ker(\theta \cdot \pi) = \ker(\pi).$$

But $\ker(\pi) = N = \mathrm{im}(1 \otimes \psi)$, by definition. Thus the proof will be complete if we show that θ is an isomorphism. We do this by explicitly constructing an inverse.

Define a map,

$$\beta : B \times M'' \to (B \otimes_S M)/N,$$

by $\beta(b, v) = b \otimes u + N$, where $\phi(u) = v$. We must check that it is well-defined. Assume that $\phi(u) = \phi(u') = v$, where $u, u' \in M$. Then $\phi(u - u') = 0$ and so

$$u - u' \in \ker(\phi) = \mathrm{im}(\psi).$$

But we have that

$$b \otimes u \equiv b \otimes u' \pmod{N}$$

and β is well-defined. On the other hand, β is clearly bilinear and S-balanced. By the universal property of the tensor product there exists a homomorphism of R-T-bimodules,

$$\bar{\beta} : B \otimes_S M'' \to (B \otimes_S M)/N.$$

A straightforward calculation shows that θ and β are inverse to each other.

Of course there is a version of Theorem 4.6 with B tensoring on the right, and we shall use it whenever necessary without any further comment. The same applies to the next result.

4.7 COROLLARY. *Let S be a K-algebra and R a subalgebra of S. Suppose that $I \subseteq S$ is an S-R-bimodule and that M is a right S-module. Then*

$$M \otimes_S S/I \cong M/MI$$

as right R-modules.

PROOF: Tensoring the sequence $I \xrightarrow{\theta} S \to S/I \to 0$ with M over S, and using Theorem 4.6, we obtain the exact sequence

$$M \otimes_S I \xrightarrow{1 \otimes \theta} M \otimes_S S \to M \otimes S/I \to 0.$$

But $M \otimes_S S \cong M$ by Proposition 4.1. The image of the composition of $1 \otimes \theta$ with this isomorphism is MI. Note that this is a right R-submodule of M. Thus,

$$M \otimes_S S/I \cong M/MI$$

as right R-modules.

An important warning: injective maps need *not* be preserved by tensor products. Consequently, tensor products do not always preserve short exact sequences. An example is given in Exercise 6.6. See also Exercise 6.5.

5. LOCALIZATION.

It is time for our first application of the tensor product to the Weyl algebras. In it we generalize a result proved in Ch.10, §3. Let p be a non-zero polynomial in $K[X]$ and consider the subset $K[X, p^{-1}]$ of all rational functions whose denominator is a power of p. We have proved that this is a holonomic left A_n-module in Theorem 10.3.2.

Let M be a $K[X]$-module and put

$$M[p^{-1}] = K[X, p^{-1}] \otimes_{K[X]} M.$$

Since $K[X]$ is commutative, this is a $K[X]$-module. Note that every element of $M[p^{-1}]$ can be written in the form $p^{-k} \otimes u$, for some $u \in M$ and $k \geq 0$, but this form is *not* unique. This poses the question: when is $p^{-k} \otimes u = 0$?

5.1 PROPOSITION. *Let M be a $K[X]$-module. Suppose that $u \in M$ satisfies $1 \otimes u = 0$ in $M[p^{-1}]$. Then there exists an integer $k \geq 0$ such that $p^k u = 0$.*

PROOF: There exist an index set $\mathcal{I}$ and a surjective map

$$K[X]^{\mathcal{I}} \xrightarrow{\theta} M,$$

where $K[X]^{\mathcal{I}}$ denotes the direct sum of copies of $K[X]$ indexed by $\mathcal{I}$: a free $K[X]$-module. Let F be the kernel of this map. Tensoring by $K[X, p^{-1}]$ over $K[X]$, we get an exact sequence,

$$(5.2) \qquad K[X, p^{-1}] \otimes F \to K[X, p^{-1}] \otimes K[X]^{\mathcal{I}} \xrightarrow{1 \otimes \theta} K[X, p^{-1}] \otimes M \to 0.$$

By Theorem 4.2,

$$K[X, p^{-1}] \otimes_{K[X]} K[X]^{\mathcal{I}} \cong K[X, p^{-1}]^{\mathcal{I}}.$$

It is easy to check that the image of the first map of (5.2) in $K[X, p^{-1}]^{\mathcal{I}}$ is $K[X, p^{-1}]F$, the submodule generated by F. Therefore

$$M[p^{-1}] \cong K[X, p^{-1}]^{\mathcal{I}} / K[X, p^{-1}]F$$

as $K[X]$-modules. Choose $v \in K[X]^{\mathcal{I}}$ such that $\theta(v) = u$. Thus

$$(1 \otimes \theta)(1 \otimes v) = 1 \otimes u = 0.$$

Identifying $1 \otimes v$ with v in $K[X, p^{-1}]^{\mathcal{I}}$, we have that $v \in K[X, p^{-1}]F$. Hence there exists $k \geq 0$ such that $p^k v \in F$; from which we conclude that $p^k u = 0$ in M.

Now assume that M is a left A_n-module. Then $M[p^{-1}]$ is a $K[X]$-module, and we want to turn it into a left A_n-module. We do that by defining the action of ∂_i as follows

$$\partial_i \cdot (p^{-k} \otimes u) = p^{-k-1} \otimes (p\partial_i \cdot u - k\partial_i(p)u).$$

It is not immediately clear that this formula gives a well-defined action, because an element of $M[p^{-1}]$ does not have a unique representation in the form $p^{-k} \otimes u$. However, applying Proposition 5.1 one can show that it is indeed well-defined; the details are left to the reader. It is also easy to check that the relations of A_n are preserved by this action. Hence $M[p^{-1}]$ is a left A_n-module by Appendix 1. It is called the *localization* of M at p. The name comes from the way this construction is used in algebraic geometry.

If we also assume that M has a good filtration Γ, then we can construct a filtration Ω for $M[p^{-1}]$. If m is the degree of p, put

(5.3) $$\Omega_k = \{p^{-k} \otimes u : u \in \Gamma_{(m+1)k}\}.$$

To check that this is a filtration of $M[p^{-1}]$, proceed as in the proof of Theorem 10.3.2.

5.4 THEOREM. *If M is a holonomic A_n-module, then so is $M[p^{-1}]$.*

PROOF: We have that

$$\dim_K \Omega_k \leq \dim_K \Gamma_{(m+1)k}.$$

Let $\chi(t, M, \Gamma)$ be the Hilbert polynomial of Γ. Then for $k \gg 0$,

$$\dim_K \Omega_k \leq \chi(k(m+1), M, \Gamma)$$

a polynomial of degree n. Hence $M[p^{-1}]$ is holonomic by Lemma 10.3.1.

6. EXERCISES

6.1 Let $p, q \in \mathbb{Z}$, and let m be their greatest common divisor. Show that $\mathbb{Z}_p \otimes_{\mathbb{Z}} \mathbb{Z}_q \cong \mathbb{Z}_m$.

6.2 Prove the following isomorphisms of $\mathbb{Z}$-modules:

(1) $\mathbb{Q} \otimes_{\mathbb{Z}} \mathbb{Q} \cong \mathbb{Q}$.

(2) $\mathbb{Q} \otimes_{\mathbb{Z}} (\mathbb{Q}/\mathbb{Z}) \cong (\mathbb{Q}/\mathbb{Z}) \otimes_{\mathbb{Z}} (\mathbb{Q}/\mathbb{Z})$.

6.3 Let R be a ring. Show that $R^n \otimes_R R^m \cong R^{nm}$.

6.4 Prove Propositions 4.4 and 4.5 in detail.

6.5 Let R be a K-algebra and let $k \geq 0$ be an integer. Show that if $N \to M$ is an injective map of left R-modules, then so is

$$R^k \otimes_R N \to R^k \otimes_R M.$$

6.6 Let $\phi : \mathbb{Z}_2 \to \mathbb{Z}_4$ be the injective homomorphism of $\mathbb{Z}$-modules defined by $\phi(1 + 2\mathbb{Z}) = 2 + 4\mathbb{Z}$. Show that:

(1) $\mathbb{Z}_2 \otimes_\mathbb{Z} \mathbb{Z}_2 \cong \mathbb{Z}_2$.

(2) $\mathbb{Z}_2 \otimes_\mathbb{Z} \mathbb{Z}_4 \cong \mathbb{Z}_2$.

(3) The map $\mathbb{Z}_2 \to \mathbb{Z}_2$ induced by tensoring ϕ by $\mathbb{Z}_2$ over $\mathbb{Z}$ is identically zero.

Conclude that the tensor product by $\mathbb{Z}_2$ does not preserve injectivity.

6.7 Show that for any non-zero $p \in K[X]$, there exist A_n-module isomorphisms:

(1) $K[X, p^{-1}] \otimes_{K[X]} K[X] \cong K[X, p^{-1}]$;

(2) $K[X, p^{-1}] \otimes_{K[X]} K[\partial_1, \ldots, \partial_n] = 0$.

CHAPTER 13
EXTERNAL PRODUCTS

In this chapter we discuss the simplest operation to be performed on modules over the Weyl algebra: the external product. We also discuss the computation of the dimensions of the external product from the dimension of its factors. In particular, we will show that the external product of holonomic modules is itself holonomic.

1. EXTERNAL PRODUCTS OF ALGEBRAS.

We begin with a general definition. Let A,B be K-algebras. The tensor product $A \otimes_K B$ is a K-vector space, on which we define a multiplication. For $a, a' \in A$ and $b, b' \in B$, let

$$(a \otimes b)(a' \otimes b') = aa' \otimes bb'.$$

It is routine to check that $A \otimes_K B$ with this product is a K-algebra. This is called the *external product* of A and B. Since we use this construction very often, it is convenient to have a special notation for it: $A \widehat{\otimes} B$.

If we apply the construction to polynomial rings or to the Weyl algebra we do not get anything new: that is the secret of its power. The key result is the following theorem.

1.1 THEOREM. *Let R be a K-algebra and let A and B be subalgebras of R. Suppose that*

(1) $R = AB$,

(2) $[A, B] = 0$,

(3) *there exist K-bases $\{a_i : i \in \mathbb{N}\}$ and $\{b_j : j \in \mathbb{N}\}$ of A and B, respectively, such that $\{a_i b_j : i, j \in \mathbb{N}\}$ is a K-basis of R.*

Then $R \cong A \widehat{\otimes} B$.

PROOF: Define a map $\phi : A \times B \to R$ by $\phi(a, b) = ab$. It is clearly K-bilinear and K-balanced. Therefore, there exists a map $\Phi : A \otimes_K B \to R$ such that $\Phi(a \otimes b) = ab$. This is called the *multiplication map*.

Let $a, a' \in A$ and $b, b' \in B$. Then

$$\Phi((a \otimes b)(a' \otimes b')) = \Phi(aa' \otimes bb') = aa'bb'.$$

On the other hand,

$$\Phi(a \otimes b)\Phi(a' \otimes b') = aba'b'.$$

Thus (2) implies that Φ is a K-algebra homomorphism.

Since, from (1), every element of R is a linear combination of monomials ab, with $a \in A$ and $b \in B$, we conclude that Φ is surjective. The injectivity is a consequence of (3) and Corollary 12.4.3.

A carefully chosen notation makes the application of Theorem 1.1 to polynomial rings and the Weyl algebra almost tautological. Our choice is the following. Let $K[X] = K[x_1, \ldots, x_n]$ and $K[Y] = K[y_1, \ldots, y_m]$ be polynomial rings. Write $K[X, Y]$ for the polynomial ring on the x's and y's. Then A_n will be the Weyl algebra generated by the x's and ∂_x's, and A_m the Weyl algebra generated by the y's and ∂_y's. Both are subalgebras of A_{m+n}, the Weyl algebra generated by the x's, y's and their derivatives. We shall retain this notation for the rest of the chapter.

1.2 COROLLARY. *The following isomorphisms are induced by the multiplication map:*

(1) $K[X] \widehat{\otimes} K[Y] \cong K[X, Y]$.

(2) $A_m \widehat{\otimes} A_n \cong A_{m+n}$.

The isomorphisms defined in the corollary are so natural that we shall take them to be equalities. Thus, from now on, we shall use without further comment that $A_m \widehat{\otimes} A_n = A_{m+n}$. In other words if $a \in A_n$ and $b \in A_m$, we will identify the monomial ab with $a \otimes b$. Similarly for polynomial rings.

2. EXTERNAL PRODUCTS OF MODULES.

Once again we turn to the general situation. Let A, B be K-algebras. Suppose that M is a left A-module and that N is a left B-module. Then we may turn the K-vector space $M \otimes_K N$ into a left $(A \widehat{\otimes} B)$-module. The action of $a \otimes b \in A \widehat{\otimes} B$ on $u \otimes v \in M \otimes_K N$ is given by the formula

$$(a \otimes b)(u \otimes v) = au \otimes bv.$$

It is routine to check that $M \otimes_K N$ is a module for this action. We shall write $M\widehat{\otimes}N$ for this $(A\widehat{\otimes}B)$-module.

2.1 PROPOSITION. *If M and N are finitely generated modules, then so is $M\widehat{\otimes}N$.*

PROOF: Suppose that M is generated by $u_1, \ldots, u_s$ over A and that N is generated over B by $v_1, \ldots, v_t$. The elements of $M\widehat{\otimes}N$ are K-linear combinations of elements of the form $u \otimes v$, for $u \in M$, $v \in N$. But $u = \sum_1^s a_i u_i$ and $v = \sum_1^t b_j v_j$. Thus,

$$u \otimes v = \sum_{i,j}(a_i \otimes b_j)(u_i \otimes v_j).$$

Hence $M\widehat{\otimes}N$ is generated over $A\widehat{\otimes}B$ by $u_i \otimes v_j$, for $1 \le i \le s$ and $1 \le j \le t$.

Now let M be a left A_m-module and N a left A_n-module. The external product $M\widehat{\otimes}N$ is a left A_{m+n}-module, under the convention that $A_{m+n} = A_m\widehat{\otimes}A_n$. This will be used so often, that even at the risk of being pedantic, we write the action explicitly. Let $u \otimes v \in M\widehat{\otimes}N$. A monomial of A_{m+n} can be written, in only one way, in the form ab with $a \in A_m$ and $b \in A_n$. Thus,

$$(ab)(u \otimes v) = au \otimes bv.$$

The situation for polynomial rings is entirely similar. The next lemma will be very useful in calculations. We make use of these conventions in its statement and proof.

2.2 LEMMA. *Let I be a left ideal of A_m and J a left ideal of A_n. Denote by $A_{m+n}I + A_{m+n}J$ the left ideal of A_{m+n} generated by the elements of I and J. Then*

$$(A_m/I)\widehat{\otimes}(A_n/J) \cong A_{m+n}/(A_{m+n}I + A_{m+n}J)$$

as A_{m+n}-modules.

PROOF: Since every monomial in A_{m+n} may be written, uniquely, in the form ab, for $a \in A_m$, $b \in A_n$, there exists a K-linear map,

$$\psi : A_{m+n} \longrightarrow (A_m/I)\widehat{\otimes}(A_n/J)$$

given by $\psi(ab) = (a + I) \otimes (b + J)$. An easy calculation shows that this is an A_{m+n}-module homomorphism. It is clearly surjective. On the other hand, a monomial ab with $a \in I$ or $b \in J$, gives $\psi(ab) = 0$. Hence $A_{m+n}I + A_{m+n}J \subseteq \ker \psi$.

By the universal property of the tensor product, there is a K-linear map,

$$\phi: (A_m/I)\widehat{\otimes}(A_n/J) \rightarrow A_{m+n}/(A_{m+n}I + A_{m+n}J),$$

defined by $\phi((a + I) \otimes (b + J)) = ab + A_{m+n}I + A_{m+n}J$. This is a surjective map, and

$$\phi\psi(ab) = ab + A_{m+n}I + A_{m+n}J.$$

Hence $A_{m+n}I + A_{m+n}J = \ker \psi$ and ψ is an isomorphism of A_{m+n}-modules.

Although we have not used in the proof of 2.2 that ϕ is a homomorphism of A_{m+n}-modules, this is easily shown to be true.

2.3 COROLLARY. *Let I be a left A_m-module. Then*

$$(A_m/I)\widehat{\otimes}A_n \cong A_{m+n}/A_{m+n}I$$

is an isomorphism of A_{m+n}-A_n-bimodules.

PROOF: It follows from Lemma 2.2 that there exists an isomorphism of left A_{m+n}-modules,

$$\psi: A_{m+n}/A_{m+n}I \rightarrow (A_m/I)\widehat{\otimes}A_n,$$

given by $\psi(ab + A_{m+n}I) = (a + I) \otimes b$, where $a \in A_m$, $b \in A_n$. Note that since $I \subseteq A_m$, the elements of I commute with the elements of A_n. Hence $A_{m+n}/A_{m+n}I$ is a right A_n-module. So is $(A_m/I)\widehat{\otimes}A_n$, with A_n acting on the right component of the tensor product. Let $c \in A_n$. Then

$$\psi((ab)c + A_{m+n}I) = (a + I) \otimes bc = \psi(ab + A_{m+n}I)c.$$

Therefore, ψ is also a right A_n-module isomorphism.

3. GRADUATIONS AND FILTRATIONS.

We will now construct a good filtration for the external product of modules over the Weyl algebra and calculate its associated graded module. Through-out this section, let M be a finitely generated left A_m-module with good filtration $\Gamma(M) = \{\Gamma_i(M) : i \in \mathbb{N}\}$ and let N be a finitely generated left A_n-module with good filtration $\Gamma(N)$. We will proceed through a series of lemmas.

3.1 LEMMA. *The Bernstein filtration of A_{m+n} satisfies*

$$B_t(A_{m+n}) = \sum_{p+q=t} B_p(A_m)B_q(A_n).$$

PROOF: First note that

$$B_p(A_m)B_q(A_n) \subseteq B_{p+q}(A_{m+n}).$$

Now every monomial in $B_t(A_{m+n})$ may be written as a product ab, with $a \in A_m$ and $b \in A_n$. Assume that a, b have degrees p, q, respectively. Since ab has degree $p+q$, it follows that $ab \in B_p(A_m)B_q(A_n)$. But an element of $B_t(A_{m+n})$ is a sum of such monomials. Therefore, $B_t(A_{m+n}) = \sum_{p+q=t} B_p(A_m)B_q(A_n)$.

3.2 LEMMA. *The K-vector spaces*

$$\Gamma_k(M\widehat{\otimes}N) = \sum_{i+j=k} \Gamma_i(M) \otimes_K \Gamma_j(N)$$

form a good filtration of $M\widehat{\otimes}N$ as an A_{m+n}-module.

Note that we are identifying $\Gamma_i(M) \otimes_K \Gamma_j(N)$ with a subspace of $M\widehat{\otimes}N$, in the obvious way.

PROOF: Since the summation in the definition of $\Gamma_k(M\widehat{\otimes}N)$ is finite, we have by Corollary 12.4.3 that $\Gamma_k(M\widehat{\otimes}N)$ is a finite dimensional K-vector space.

Now every element of $M\widehat{\otimes}N$ is a finite linear combination of elements of the form $u \otimes v$, with $u \in M$, $v \in N$. Assume that $u \in \Gamma_i(M)$ and $v \in \Gamma_j(N)$. Then

$$u \otimes v \in \Gamma_i(M) \otimes \Gamma_j(N) \subseteq \Gamma_{i+j}(M\widehat{\otimes}N).$$

Hence, $M\widehat{\otimes}N = \bigcup_{k\geq0} \Gamma_k(M\widehat{\otimes}N)$.

Finally, since

$$(B_p(A_m)B_q(A_n))(\Gamma_i(M) \otimes_K \Gamma_j(N)) \subseteq \Gamma_{i+p}(M) \otimes_K \Gamma_{j+q}(N),$$

we conclude, using Lemma 3.1, that $B_t(A_{m+n})\Gamma_k(M\widehat{\otimes}N) = \Gamma_{k+t}$ for $t \gg 0$. Thus $\Gamma_k(M\widehat{\otimes}N)$ is a good filtration of $M\widehat{\otimes}N$.

We will now calculate the graded module with respect to this good filtration. The ring $S_n = gr^B A_n$ is a polynomial ring by Theorem 7.3.1. Using the conventions of §1 we may identify $S_m\widehat{\otimes}S_n$ with S_{m+n}. Denote the homogeneous component of degree t of S_n by $S_n(t)$. By an argument similar to that used in Lemma 3.1, we have that $S_{n+m}(t) = \bigoplus_{p+q=t} S_p(A_m)S_q(A_n)$. Finally, write $gr_k(M) = \Gamma_k(M)/\Gamma_{k-1}(M)$.

3.3 LEMMA. *There is an isomorphism of K-vector spaces*

$$gr_k(M\widehat{\otimes}N) \cong \bigoplus_{i+j=k} gr_i(M) \otimes_K gr_j(N).$$

PROOF: Consider first the K-linear map,

$$\eta_{ij} : \Gamma_i(M) \otimes_K \Gamma_j(N) \to gr_i(M) \otimes_K gr_j(N)$$

defined by $\eta_{ij}(u\otimes v) = \mu_i(u)\otimes\mu_j(v)$, where we are denoting the symbol maps of both M and N by μ. It is easy to see that η_{ij} is surjective and that its kernel is the subspace

$$\Gamma_{i-1}(M) \otimes_K \Gamma_j(N) + \Gamma_i(M) \otimes_K \Gamma_{j-1}(N).$$

Putting these maps together, we get a linear map

$$\eta : \bigoplus_{i+j=k} (\Gamma_i(M) \otimes \Gamma_j(N)) \to \bigoplus_{i+j=k} (gr_i(M) \otimes gr_j(N)).$$

Note that the canonical projection gives a map

$$\bigoplus_{i+j=k} (\Gamma_i(M) \otimes \Gamma_j(N)) \to \Gamma_k(M\widehat{\otimes}N).$$

But $\ker \eta_{ij} \subseteq \Gamma_{i+j-1}(M\widehat{\otimes}N)$, thus

$$\bigoplus_{i+j=k} (\Gamma_i(M) \otimes \Gamma_j(N))\Big/ \bigoplus_{i+j=k} \ker \eta_{ij} \to \bigoplus_{i+j=k} (gr_i(M) \otimes gr_j(N)).$$

factors through

$$\Gamma_k(M\widehat{\otimes}N)/\Gamma_{k-1}(M\widehat{\otimes}N) \to \bigoplus_{i+j=k}(gr_i(M) \otimes gr_j(N)).$$

Since the former map is bijective, so is the latter, which gives the required isomorphism.

Taking direct sums for $k \geq 0$, and using Lemma 3.3, we conclude that there exists an isomorphism of K-vector spaces,

$$\theta : gr(M\widehat{\otimes}N) \to gr(M)\widehat{\otimes}gr(N).$$

Since $M\widehat{\otimes}N$ is an A_{m+n}-module, $gr(M\widehat{\otimes}N)$ is an S_{m+n}-module. On the other hand, since $S_{m+n} = S_m\widehat{\otimes}S_n$, it follows that $gr(M)\widehat{\otimes}gr(N)$ is an S_{m+n}-module.

3.4 THEOREM. *The linear map θ is an isomorphism of S_{m+n}-modules.*

PROOF: We have only to check that θ is compatible with the action of S_{m+n}. Choose a monomial of S_{m+n}, and write it in the form fg, with $f \in S_m(p)$, $g \in S_n(q)$. There exist operators $a, b \in A_{m+n}$, such that

$$f = \sigma_p(a) \text{ and } g = \sigma_q(b).$$

In particular, $a \in A_m$ and $b \in A_n$. Now let $u \otimes v \in \Gamma_i(M) \otimes \Gamma_j(N)$, with $i+j = k$. Then

$$(fg)\mu_k(u \otimes v) = \mu_{k+p+q}((ab)(u \otimes v)).$$

But $(ab)(u \otimes v) = au \otimes bv$; therefore

$$(fg)\mu_k(u \otimes v) = \mu_{k+p+q}(au \otimes bv).$$

The latter is mapped onto $\mu_{i+p}(au) \otimes \mu_{j+q}(bv)$ by θ. Since this may be rewritten as $(fg)(\mu_i(u) \otimes \mu_j(v))$, we have

$$\theta((fg)\mu_k(u \otimes v)) = (fg)(\mu_i(u) \otimes \mu_j(v))$$

which completes the proof.

4. DIMENSIONS AND MULTIPLICITIES

We are now ready to calculate the dimension and multiplicity of an external product.

4.1 THEOREM. *Let M be a finitely generated left A_m-module and N a finitely generated left A_n-module. Then*

(1) $d(M \widehat{\otimes} N) = d(M) + d(N)$,

(2) $m(M \widehat{\otimes} N) \leq m(M) m(N)$.

PROOF: We retain the notation of §3. It follows from Lemma 3.3 that

$$ dim_K \Gamma_k(M \widehat{\otimes} N) = \sum_{r=0}^{k} \sum_{i+j=r} dim_K gr_i(M) dim_K gr_j(N). $$

Therefore,

$$ dim_K \Gamma_k(M \widehat{\otimes} N) \leq \sum_{i=0}^{k} dim_K gr_i(M) \sum_{j=0}^{k} dim_K gr_j(N). $$

From this inequality and Lemma 3.2 we have that

$$ dim_K \Gamma_k(M \widehat{\otimes} N) \leq dim_K \Gamma_k(M) dim_K \Gamma_k(N) \leq dim_K \Gamma_{2k}(M \widehat{\otimes} N). $$

Assuming that $k \gg 0$, and using the Hilbert polynomials of the corresponding filtrations we get that

$$ \chi(k, M \widehat{\otimes} N) \leq \chi(k, M) \chi(k, N) \leq \chi(2k, M \widehat{\otimes} N). $$

Since this holds for all large values of k, both (1) and (2) immediately follow.

The corollary is an easy consequence of the theorem.

4.2 COROLLARY. *Let M be a holonomic A_m-module and N a holonomic A_n-module. Then $M \widehat{\otimes} N$ is a holonomic A_{m+n}-module.*

5. EXERCISES

5.1 Let B_n be the ring of differential operators of the rational function field $K(X) = K(x_1, \ldots, x_n)$. Is it true that $B_n \widehat{\otimes} B_m \cong B_{n+m}$ as K-algebras?

5.2 Show that the multiplication map induces an isomorphism of A_{m+n}-modules,

$$ K[X, Y] \cong K[X] \widehat{\otimes} K[Y]. $$

5.3 Let $p \in K[X]$ and $q \in K[Y]$. Show that there exists an isomorphism of A_{m+n}-modules,

$$K[X, Y, (pq)^{-1}] \cong K[X, p^{-1}] \widehat{\otimes} K[Y, q^{-1}].$$

5.4 Let σ, σ' be automorphisms of A_m and A_n, respectively. Let M be a left A_m-module and N be a left A_n-module.

(1) Show that there exists a unique automorphism θ of A_{m+n} whose restriction to A_m is σ and whose restriction to A_n is σ'.

(2) Show that

$$M_\sigma \widehat{\otimes} N_{\sigma'} \cong (M \widehat{\otimes} N)_\theta.$$

5.5 Let M be a left A_m-module and N a left A_n-module. Show that the Fourier Transform satisfies

$$(M \widehat{\otimes} N)_{\mathcal{F}} \cong M_{\mathcal{F}} \widehat{\otimes} N_{\mathcal{F}}$$

5.6 Let M be a holonomic A_m-module and N a holonomic A_n-module. Show that if M and N have multiplicity 1, then $M \widehat{\otimes} N$ is a simple holonomic A_{m+n}-module.

5.7 In this exercise we use the results of Ch. 11. Let M be a left A_m-module and N a left A_n-module. Show that $Ch(M \widehat{\otimes} N) = Ch(M) \times Ch(N)$.

5.8 Let $1 \le i \le n$. Suppose that d_i is an operator of A_n of degree ≥ 1 which is a linear combination of monomials in x_i and ∂_i. Let J be the left ideal generated by $d_1, \ldots, d_n$. Show that A_n/J is a holonomic module over A_n. Hint: $A_n/J \cong (A_1/A_1 d_1) \widehat{\otimes} \ldots \widehat{\otimes} (A_1/A_1 d_n)$.

CHAPTER 14

INVERSE IMAGES

In this chapter we show how to change the ring of scalars of a module over a Weyl algebra. This is very important and leads to the construction of many new modules. As we shall see later, holonomic modules are preserved under this construction.

1. CHANGE OF RINGS

We start with a general construction for rings and modules. Let R, S be rings, and let $\phi : R \to S$ be a ring homomorphism. If M is a left R-module, then we may use ϕ to turn it into a left S-module. This is called *changing the base ring*. The key to the construction is that S may be considered as a right R-module in the following way. Let $r \in R$, $s \in S$; the right action of R on S is defined by

$$s \star r = s\phi(r).$$

The juxtaposition on the right hand side of the equation denotes multiplication in the ring S.

With this in mind, we may consider S as an S-R-bimodule. We are now allowed to take the tensor product $S \otimes_R M$ which is a left S-module. Thus starting with the left R-module M we have constructed the left S-module $S \otimes_R M$. If it is necessary to call attention to the homomorphism ϕ, we will write $S \otimes_\phi M$.

As an application, let us rephrase the twisting construction of Ch. 5, §2 in terms of change of rings.

1.1 LEMMA. *Let R be a ring and σ an automorphism of R. Let M be a left R-module. Then*

$$M_\sigma \cong R \otimes_{\sigma^{-1}} M.$$

PROOF: Let $a \in R$. The left action on M_σ is given by $a \bullet u = \sigma(a)u$, where $u \in M_\sigma$; and the right action on R by $s \star a = s\sigma^{-1}(a)$, where $s \in R$. Consider the map

$$\phi : R \times M \to M_\sigma$$

defined by $\phi(a, u) = a \bullet u$. Since

$$\phi(a, bu) = a \bullet bu = (\sigma(a)b)u$$

is equal to $(a\sigma^{-1}(b)) \bullet u = \phi(a \star b, u)$, the map is balanced. It is also bilinear. Thus, by the universal property of the tensor product, there exists a homomorphism

$$\overline{\phi} : R \otimes_{\sigma^{-1}} M \to M_\sigma.$$

Since $M_\sigma = M$ as abelian groups, the map $\overline{\phi}$ is surjective.

Let $v \in R \otimes_{\sigma^{-1}} M$. There exists $u \in M$ such that $v = 1 \otimes u$. Hence

$$0 = \overline{\phi}(v) = u,$$

which implies that $u = 0$. Thus $v = 0$ and $\overline{\phi}$ is injective. Therefore, $\overline{\phi}$ is an isomorphism of R- modules.

We shall now apply the change of rings construction to polynomial maps. We often use the results of Ch. 4, §1 in the coming chapters, so you may wish to read that section again before you proceed. A good notation will be of great help: we shall abide by the conventions we are about to make right to the end of the book.

1.2 NOTATION. *Put $X = K^n$. The polynomial ring $K[x_1, \ldots, x_n]$ will be denoted by $K[X]$; and the Weyl algebra generated by the x's and ∂_x's by A_n. The n-tuple $(x_1, \ldots, x_n)$ will be denoted by X. Similar conventions will hold for $Y = K^m$ and $Z = K^r$, with polynomial rings $K[Y]$ and $K[Z]$ and Weyl algebras A_m and A_r. The space K^{m+n} equals the cartesian product of X and Y and will be denoted by $X \times Y$. For the polynomial ring in the x's and y's we shall write $K[X, Y]$. The $(n + m)$-th Weyl algebra A_{m+n} will always denote the algebra generated by $K[X, Y]$, ∂_x's and ∂_y's. Note that with these conventions $A_{m+n} = A_m \widehat{\otimes} A_n$.*

Although these conventions may cause some initial confusion, they pay off later on by making the formulae easier to digest than they would otherwise be.

Let $F : X \to Y$ be a polynomial map. Its comorphism $F^\sharp : K[Y] \to K[X]$ is an algebra homomorphism. If M is a $K[Y]$-module, we may use $F^\sharp$ to construct the $K[X]$-module $K[X] \otimes_{K[Y]} M$. It is called the *inverse image* of M by the polynomial map F, and denoted by F^*M. Let us consider some examples.

Let $F : X \longrightarrow X$ be a polynomial isomorphism, with inverse G. The comorphism $F^\sharp$ is an automorphism of $K[X]$ with inverse $G^\sharp$. Let M be a $K[X]$-module. By Lemma 1.1, its inverse image F^*M is isomorphic to $M_{G^\sharp}$. Consequently, if $u \in F^*M$ and $h \in K[X]$, then $hu = G^\sharp(h)u$.

The second example is the projection $\pi : X \times Y \longrightarrow Y$ onto the second coordinate: $\pi(X, Y) = Y$. The comorphism $\pi^\sharp$ maps a polynomial $p \in K[Y]$ to itself, but considered as an element of $K[X, Y]$. Let M be a module over $K[Y]$, then

$$\pi^*M = K[X, Y] \otimes_{K[Y]} M.$$

A monomial in $K[X, Y]$ may be written in the form pq, with $p \in K[X]$ and $q \in K[Y]$. If $u \in M$ then

$$pq \otimes u = p \otimes qu$$

as elements of π^*M. Using this identity we may construct an isomorphism,

$$\pi^*M \cong K[X]\widehat{\otimes}M,$$

of $K[X, Y]$-modules.

2. INVERSE IMAGES.

Let $F : X \longrightarrow Y$ be a polynomial map and M a left A_m-module. Since $Y = K^m$, it follows that M is in particular a $K[Y]$-module. Thus we may compute the inverse image of M by F defined in §1:

$$F^*(M) = K[X] \otimes_{K[Y]} M$$

is a module over $K[X]$. We want to make this module into an A_n-module.

At first sight this construction may seem very puzzling. Why do *change of rings* as if we had only a module over a polynomial ring and then turn it back into a module over a Weyl algebra? Why not do the change of rings at the Weyl algebra level? First of all, one is allowed to do change of rings at the level of Weyl algebras, since the construction works for any ring. However, in doing that one is limited by the fact that Weyl algebras are simple rings; and homomorphisms of simple rings are necessarily injective. A more satisfactory answer is that this is really a construction in algebraic geometry, that we are lifting to the realm of noncommutative rings; see [Hartshorne **77**, Ch. 2, §5]. Anyhow, the efficacy of the construction is its best justification, as we shall see later on.

We know how a polynomial of $K[X]$ acts on $F^*(M)$. We now give the recipe for the action of ∂_{x_i}. Let

$$q \otimes u \in F^*(M),$$

where $q \in K[X]$ and $u \in M$. Let $F_1, \ldots, F_m$ be the coordinate functions of F. The action of ∂_{x_i} is defined by

$$(2.1) \qquad \partial_{x_i}(q \otimes u) = \partial_{x_i}(q) \otimes u + \sum_{k=1}^{m} q\partial_{x_i}(F_k) \otimes \partial_{y_k} u$$

for $i = 1, \ldots, n$.

The x's and ∂_x's generate A_n, and we know how they act on $F^*(M)$. According to Appendix 1, these formulae will make an A_n-module of $F^*(M)$ if they are compatible with the relations satisfied by the generators. The relations are

$$[\partial_{x_j}, x_i] = \delta_{ij}1,$$
$$[x_i, x_j] = [\partial_{x_i}, \partial_{x_j}] = 0,$$

for $1 \le i, j \le n$. Let us carefully check that the first of these relations is compatible with the actions of x's and ∂_x's defined above.

Let $w \in F^*(M)$. We want to show that

$$[\partial_{x_j}, x_i]w = \delta_{ij}w.$$

Clearly it is enough to check this when w is of the form $q \otimes u$, for $q \in K[X]$ and $u \in M$. Let us begin by calculating $\partial_{x_j}(x_i(q \otimes u))$. We do this by applying (2.1) to $x_iq \otimes u$, which gives

$$(2.2) \qquad \partial_{x_j}(x_iq) \otimes u + \sum_{1}^{m} x_iq\partial_{x_j}(F_k) \otimes \partial_{y_k} u.$$

Using Leibniz's rule and the fact that $\partial_{x_j}(x_i) = \delta_{ij}$, we may rewrite (2.2) in the form

$$\delta_{ij} q \otimes u + x_i \left(\partial_{x_j}(q) \otimes u + \sum_1^m (q \partial_{x_j}(F_k) \otimes \partial_{y_k} u) \right).$$

Note that the expression in brackets equals $\partial_{x_j}(q \otimes u)$. Thus we have shown that

$$\partial_{x_j}(x_i(q \otimes u)) = \delta_{ij} q \otimes u + x_i \partial_{x_j}(q \otimes u).$$

But this is equivalent to

$$[\partial_{x_j}, x_i](q \otimes u) = \delta_{ij}(q \otimes u),$$

which is what we wanted to prove.

The proof that the other relations are compatible with these actions is an exercise for the reader. We will calculate some concrete examples in the next section.

3. PROJECTIONS.

Let us return to the projection $\pi : X \times Y \to Y$ of §1. Let M be a left A_m-module. We have already seen that, as a $K[X, Y]$-module, the inverse image $\pi^* M$ is isomorphic to $K[X] \widehat{\otimes} M$. However, $K[X]$ is an A_n-module and M is an A_m-module. So $K[X] \widehat{\otimes} M$ has a natural A_{m+n}-module structure. We want to show that this structure coincides with the one determined by (2.1).

It is enough to check how ∂_{x_i} and ∂_{y_j} act on $K[X] \widehat{\otimes} M$. For this we return to the definitions. The $K[X, Y]$-module isomorphism between $\pi^*(M)$ and $K[X] \widehat{\otimes} M$ maps $q_\alpha x^\alpha \otimes u$ to $x^\alpha \otimes q_\alpha u$, where $\alpha \in \mathbb{N}^n$ and $q_\alpha \in K[Y]$. We must show that this is compatible with the action of the derivatives.

Consider first the action of ∂_{y_j}, as defined by (2.1),

$$\partial_{y_j}(q_\alpha x^\alpha \otimes u) = \partial_{y_j}(q_\alpha x^\alpha) \otimes u + \sum_1^m (\partial_{y_j}(y_k) q_\alpha x^\alpha \otimes \partial_{y_k} u).$$

Since $\partial_{y_j}(x^\alpha) = 0$ and $\partial_{y_j}(y_k) = \delta_{jk}$, we obtain

$$\partial_{y_j}(q_\alpha x^\alpha \otimes u) = x^\alpha \partial_{y_j}(q_\alpha) \otimes u + q_\alpha x^\alpha \otimes \partial_{y_j} u.$$

Using Leibniz's rule, this may be rewritten as

$$\partial_{y_j}(q_\alpha x^\alpha \otimes u) = x^\alpha \otimes \partial_{y_j}(q_\alpha u).$$

Summing up: if we identify $\pi^*(M)$ with $K[X]\widehat{\otimes}M$, then ∂_{y_j} acts only on M, in the usual way.

Consider now the action of ∂_{x_i}. From the definition, we have that

$$\partial_{x_i}(q_\alpha x^\alpha \otimes u) = \partial_{x_i}(q_\alpha x^\alpha) \otimes u + \sum_1^m \partial_{x_i}(y_k)q_\alpha x^\alpha \otimes \partial_{x_k} u.$$

Since $\partial_{x_i}(y_k) = \partial_{x_i}(q_\alpha) = 0$, for all k and all α, we end up with

$$\partial_{x_i}(q_\alpha x^\alpha \otimes u) = q_\alpha \partial_{x_i}(x^\alpha) \otimes u.$$

Thus, under the previous identification, ∂_{x_i} acts only on $K[X]$.

We may sum up our calculations as follows. If the module $\pi^*(M)$ is identified with $K[X]\widehat{\otimes}M$ then the x's and ∂_x's act only on the first factor $K[X]$, and the y's and ∂_y's act only on the second factor M. Besides, the actions are the natural ones.

We may use this description to calculate the dimension of an inverse image.

3.1 THEOREM. *Let M be a finitely generated left A_m-module and π the projection defined above. Then:*

(1) *$\pi^* M$ is a finitely generated A_{m+n}-module.*

(2) *$d(\pi^* M) = n + d(M)$.*

(3) *$m(\pi^* M) \le m(M)$.*

PROOF: (1) follows from Proposition 13.2.1; whilst (2) and (3) are consequences of Theorem 13.4.1 since, as an A_n-module, $K[X]$ has dimension n and multiplicity 1; see Ch. 9, §2.

3.2 COROLLARY. *Let M be a holonomic A_m-module. Then $\pi^* M$ is a holonomic A_{m+n}-module.*

4. EXERCISES

4.1 Let R, S, T be rings and $\phi : R \to S$ and $\psi : S \to T$ be ring homomorphisms. If M is a left R-module, show that

$$T \otimes_\psi S \otimes_\phi M \cong T \otimes_{\psi\phi} M.$$

4.2 Let $F : K \to K$ be the polynomial map defined by $F(x) = x^2$. Let M be the algebra A_1 considered as a left module over itself. Show that F^*M is not finitely generated over A_1.

4.3 Let σ be an automorphism of A_n. Suppose that the restriction θ of σ to $K[X]$ is an automorphism of this ring. Put $F = \theta_\sharp$ for the polynomial map determined by θ. Let M be a left A_n-module. Is it true that

$$F^*M \cong M_\sigma$$

as A_n-modules?

4.4 Let M and N be left A_n-modules. In particular these are modules over $K[X]$. Hence the tensor product $M \otimes_{K[X]} N$ is a well-defined $K[X]$-module. Define the action of ∂_{x_i} on $u \otimes v \in M \otimes_{K[X]} N$ by the formula

$$\partial_{x_i}(u \otimes v) = \partial_{x_i}u \otimes v + u \otimes \partial_{x_i}v.$$

(1) Proceeding as in §2, show that this action makes $M \otimes_{K[X]} N$ into an A_n-module.

(2) Give an example of two finitely generated A_n-modules M and N such that $M \otimes_{K[X]} N$ is not finitely generated over A_n.

4.5 Let $F : \mathbb{C} \to \mathbb{C}$ be the map defined by $y = F(x) = x^m$, where $m \geq 2$ is an integer. Let δ be the Dirac microfunction; see Ch. 6, §3. The purpose of this exercise is to show that the inverse image $F^*(A_1(\mathbb{C})\delta)$ is isomorphic to the $A_1(\mathbb{C})$-module generated by δ^m, the m-th derivative of δ.

(1) Show that x^m and $x\partial_x + m$ annihilate $1 \otimes \delta$.

(2) Show by induction that there exist non-zero complex numbers c_{pq} such that

$$\partial_x^{mp+q}(1 \otimes \delta) = c_{pq}x^{m-q} \otimes \partial_y^{p+1}\delta,$$

for $q = 0, 1, \ldots, m$.

(3) Using (2) show that

$$1 \otimes \partial_y^{p+1}\delta = \frac{1}{c_{pm}}\partial_x^{(p+1)m}(1 \otimes \delta).$$

(4) Conclude from (3) that $1 \otimes \delta$ generates $F^*(A_1(\mathbb{C})\delta)$.

(5) By (1), (3) and Exercise 6.4.10, $F^*(A_1(\mathbb{C})\delta)$ is a homomorphic image of $A_1(\mathbb{C})\delta^m$. Since the latter is irreducible, we have the desired isomorphism.

In the language of microfunctions we have proved that the inverse image of δ under F is δ^m.

CHAPTER 15

EMBEDDINGS

In the previous chapter we defined the inverse image of a module and calculated a formula for its dimension and multiplicity under projections. In this chapter we turn to embeddings. As we shall see, the behaviour of the inverse image under embeddings is a lot less regular than under projections.

1. The standard embedding

In this chapter we shall retain the notation of 14.1.2. We begin by considering the embedding $\iota : X \to X \times Y$ defined by $\iota(X) = (X, 0)$, where 0 denotes the origin of Y. We will call ι the *standard embedding*. The comorphism $\iota^{\sharp} : K[X, Y] \to K[X]$ is defined by $\iota^{\sharp}(g(X, Y)) = g(X, 0)$. It may be used to make $K[X]$ into a $K[X, Y]$-module; as such, $K[X]$ is generated by 1, since

$$g(X, Y) \cdot 1 = \iota^{\sharp}(g(X, Y)) \cdot 1 = g(X, 0).$$

The annihilator of 1 in $K[X, Y]$ is the ideal generated by $y_1, \ldots, y_m$. Denoting this ideal by (Y) we have that $K[X] \cong K[X, Y]/(Y)$ as $K[X, Y]$-modules.

Now consider a left A_{m+n}-module M. Since $A_n \subseteq A_{m+n}$, the A_{m+n}-module M is also an A_n-module. The elements of A_n commute with the variables $y_1, \ldots, y_m$, thus $(Y)M$ is an A_n-submodule of M. Hence $M/(Y)M$ is an A_n-module. Note however that it is *not* an A_{m+n}-submodule. We want to show that

$$\iota^* M = K[X] \otimes_{K[X, Y]} M \cong M/(Y)M$$

as A_n-modules.

If $u \in M$, let $\bar{u}$ denote its image in $M/(Y)M$. Define the map

$$\phi : \iota^* M \to M/(Y)M$$

by $\phi(q \otimes u) = q\bar{u}$. It is a homomorphism of $K[X]$-modules. We want to show that the action of the derivations $\partial_{x_1}, \ldots, \partial_{x_n}$ is compatible with ϕ. We have that

$$\partial_{x_i}(q \otimes u) = \frac{\partial q}{\partial x_i} \otimes u + \sum_{k=1}^{n} q \frac{\partial x_k}{\partial x_i} \otimes \partial_{x_k} u.$$

and so

$$\partial_{x_i}(q \otimes u) = \frac{\partial q}{\partial x_i} \otimes u + q \otimes \partial_{x_i} u.$$

The right hand side is mapped by ϕ onto

$$\frac{\partial q}{\partial x_i} \overline{u} + q \overline{\partial_{x_i} u},$$

which is equal to

$$\partial_{x_i}(q\overline{u}) = \partial_{x_i}(\phi(q \otimes u)).$$

Thus ϕ is an isomorphism of A_n-modules.

Summing up: $\iota^* M \cong M/(Y)M$ as A_n-modules. Since $M/(Y)M$ is a quotient of A_n-modules, the action of x's and ∂'s is the natural one.

Let us specialize this construction to the standard embedding $\iota : K \to K^2$, which takes x to $(x, 0)$. Let M be the ring A_2 itself, considered as a left A_2-module. Applying the above results, we have that $\iota^* M$ is the quotient A_2/yA_2 taken as a left A_1-module. Note that although $M = A_2$ is a finitely generated left A_2-module, the module $\iota^* M$ is not finitely generated. In fact, $\iota^* M$ is a free A_1-module whose basis is the image of the powers of ∂_y in A_2/yA_2. Since this basis is infinite, $\iota^* M$ is not a finitely generated left A_1-module.

We conclude from this example that the inverse image under embeddings does not preserve the noetherian property. This is very important, since we have defined dimensions only for noetherian modules. In particular there can be no general formula for the dimension of an inverse image under an embedding. Despite this, holonomic modules are preserved by inverse images under embeddings. This mystifying fact will have to wait until Ch. 18 for an explanation.

It will be necessary, for future purposes, to consider a more general kind of embedding, defined as a composition of the standard embedding and a polynomial isomorphism. Before we do this we must study the behaviour of the inverse image under composition of polynomial maps.

2. Composition.

Let us first consider what happens to the change of rings construction of Ch. 14, §1 under composition. Let R, S and T be rings and $\phi : R \to S$

and $\psi : S \rightarrow T$ be ring homomorphisms. Let M be a left R-module. We may turn M into a T-module in two different ways. On the one hand, we may apply the change of rings construction twice to get $T \otimes_\psi S \otimes_\phi M$. On the other hand, the ring T is a right R-module via the homomorphism $\psi\phi$, therefore $T \otimes_{\psi\phi} M$ is a T-module, where

$$t \otimes ru = t\psi\phi(r) \otimes u$$

for $t \in T$, $r \in R$ and $u \in M$. Using Propositions 4.1 and 4.4 of Ch. 12, we conclude that these T-modules are isomorphic; thus,

$$T \otimes_\psi S \otimes_\phi M \cong T \otimes_{\psi\phi} M.$$

It is not difficult to give a direct proof of this isomorphism. The T-module $T \otimes_\psi S \otimes_\phi M$ is generated by elements of the form $t \otimes s \otimes u$, where $t \in T$, $s \in S$ and $u \in M$. But

$$t \otimes s \otimes u = t\psi(s) \otimes 1 \otimes u.$$

Define a T-module homomorphism

$$\theta : T \otimes_\psi S \otimes_\phi M \rightarrow T \otimes_{\psi\phi} M$$

by $\theta(t \otimes s \otimes u) = t\psi(s) \otimes u$. Note that if $r \in R$ then the element

$$t \otimes 1 \otimes ru = t\psi\phi(r) \otimes 1 \otimes u$$

is mapped to $t\psi\phi(r) \otimes u$. It is easy to construct an explicit inverse for θ; hence it is an isomorphism.

We will now apply this to A_n-modules. Recall that by Theorem 4.1.1, if $F : X \rightarrow Y$ and $G : Y \rightarrow Z$ are polynomial maps, then $(GF)^\sharp = F^\sharp G^\sharp$. Note the change in the order of the maps.

2.1 THEOREM. *Let* $F : X \rightarrow Y$ *and* $G : Y \rightarrow Z$ *be polynomial maps and* M *be an* A_r-*module. Then*

$$F^* G^* M \cong (GF)^* M$$

as A_n-modules.

PROOF: By definition,

$$(GF)^*M = K[X] \otimes_{K[Z]} M.$$

In this formula $K[X]$ is a right $K[Z]$-module. Recall that the action of $f \in K[Z]$ on $h \in K[X]$ is given by $h(f \cdot G \cdot F)$. On the other hand,

$$F^*G^*M = K[X] \otimes_{K[Y]} K[Y] \otimes_{K[Z]} M.$$

We have already seen that these two modules are isomorphic as $K[X]$-modules. The isomorphism θ is defined by

$$\theta(h \otimes g \otimes u) = h(g \cdot F) \otimes u.$$

We must check that θ is compatible with the action of the derivatives ∂_x. By definition,

$$\partial_{x_i}(h \otimes g \otimes u) = \frac{\partial h}{\partial x_i} \otimes (g \otimes u) + h \sum_{j=1}^m \left(\frac{\partial F_j}{\partial x_i} \otimes \partial_{y_j}(g \otimes u) \right)$$

where $F_1, \ldots, F_m$ are the coordinate functions of F. Replacing $\partial_{y_j}(g \otimes u)$ by its formula, the term

$$\sum_{j=1}^m \left(\frac{\partial F_j}{\partial x_i} \otimes \partial_{y_j}(g \otimes u) \right)$$

becomes

$$\sum_{j=1}^m \left(\frac{\partial F_j}{\partial x_i} \otimes \frac{\partial g}{\partial y_j} \otimes u \right) + \sum_{j=1}^m \sum_{k=1}^r \left(\frac{\partial F_j}{\partial x_i} \otimes g \frac{\partial G_k}{\partial y_j} \otimes \partial_{z_k} u \right).$$

Applying θ to this expression:

$$\sum_{j=1}^m \left(\frac{\partial F_j}{\partial x_i} \left(\frac{\partial g}{\partial y_j} \cdot F \right) \otimes u \right) + \sum_{j=1}^m \sum_{k=1}^r \left((g \cdot F) \frac{\partial F_j}{\partial x_i} \left(\frac{\partial G_k}{\partial y_j} \cdot F \right) \otimes \partial_{z_k} u \right),$$

which, by interchanging the summations and applying the chain rule, becomes

$$\sum_{j=1}^m \left(\frac{\partial F_j}{\partial x_i} \left(\frac{\partial g}{\partial y_j} \cdot F \right) \otimes u \right) + \sum_{k=1}^r (g \cdot F) \left(\frac{\partial(G_k \cdot F)}{\partial x_i} \otimes \partial_{z_k} u \right).$$

Substituting the last expression in $\theta(\partial_{x_i}(h \otimes g \otimes u))$, one obtains

$$(g \cdot F)\frac{\partial h}{\partial x_i} \otimes u + h\sum_{j=1}^{m}\left(\frac{\partial F_j}{\partial x_i}\left(\frac{\partial g}{\partial y_j} \cdot F\right) \otimes u\right) + h\sum_{k=1}^{r}(g \cdot F)\frac{\partial(G_k \cdot F)}{\partial x_i} \otimes \partial_{z_k} u.$$

Another application of the chain rule and Leibniz's rule to the first two terms shows that this is equal to

$$\frac{\partial h(g \cdot F)}{\partial x_i} \otimes u + h\sum_{k=1}^{r}(g \cdot F)\left(\frac{\partial(G_k \cdot F)}{\partial x_i} \otimes \partial_{z_k} u\right),$$

and hence to $\partial_{x_i}(h(g \cdot F) \otimes u)$. We have thus proved that

$$\theta(\partial_{x_i}(h \otimes g \otimes u)) = \partial_{x_i}(h(g \cdot F) \otimes u).$$

Thus θ is an isomorphism of A_n-modules.

Applying this theorem to polynomial isomorphisms we get the following corollary.

2.2 COROLLARY. *Let $F : X \to X$ be a polynomial isomorphism and G its inverse. If M is a left A_n-module, then $F^*G^*M \cong M$.*

3. EMBEDDINGS REVISITED.

In this section we study a more general type of embedding. Let $F : X \to Y$ be a polynomial map. Define a new polynomial map $j : X \to X \times Y$ by $j(X) = (X, F(X))$. This is an injective map, as one easily checks.

We shall write j as a composition of two maps, as follows. Let $\iota : X \to X \times Y$ be the standard embedding of §1. Let $G : X \times Y \to X \times Y$ be the polynomial map defined by $G(X, Y) = (X, Y + F(X))$. Then G is bijective and $j = G \cdot \iota$.

Now let M be a left A_{m+n}-module. We will calculate j^*M. By Theorem 2.1, it equals ι^*G^*M. Let us compute G^*M. First of all, by Corollary 1.3.2, there is an automorphism σ of A_{m+n} which maps x_i and ∂_{y_j} to themselves and satisfies

$$\sigma(y_j) = y_j - F_j(X),$$

$$\sigma(\partial_{x_i}) = \partial_{x_i} + \sum_{j=1}^{n}(\partial F_j/\partial x_i)\,\partial_{y_j}.$$

Note that σ restricts to an automorphism of $K[X, Y]$ which we will also call σ. This automorphism is the inverse of the comorphism $G^\sharp$. By Lemma 14.1.1,

$$G^*M \cong M_\sigma$$

as a $K[X, Y]$-module. Let ψ stand for this isomorphism; it satisfies $\psi(h \otimes u) = \sigma(h)u$, for $h \in K[X, Y]$ and $u \in M$.

3.1 THEOREM. Let M be a left A_{m+n}-module. Then $G^*M \cong M_\sigma$ as A_{m+n}-modules.

PROOF: We know that the isomorphism holds for $K[X, Y]$-modules. Let us investigate the behaviour of the action of ∂_{x_i} under ψ . By definition, $\partial_{x_i}(h \otimes u)$ equals

$$\partial_{x_i}(h) \otimes u + h \otimes \partial_{x_i} u + \sum_{1}^{n} h \frac{\partial(F_j)}{\partial x_i} \otimes \partial_{y_j} u.$$

Applying ψ to this formula, we get

(3.2) $$\sigma(\partial_{x_i}(h))u + \sigma(h)\partial_{x_i}u + \sum_{1}^{n} \sigma(h \partial F_j / \partial x_i)\partial_{y_j} u.$$

Since σ leaves the elements of $K[X]$ unchanged, we have that $\sigma(F_j) = F_j$. Thus (3.2) is equal to

$$\sigma(\partial_{x_i}(h))u + \sigma(h)\sigma(\partial_{x_i})u.$$

But $[\partial_{x_i}, h] = \partial_{x_i}(h)$. Hence

$$\sigma(\partial_{x_i}(h)) + \sigma(h)\sigma(\partial_{x_i}) = \sigma(\partial_{x_i})\sigma(h),$$

from which we deduce that

$$\psi(\partial_{x_i}(h \otimes u)) = \sigma(\partial_{x_i})\sigma(h)u.$$

Thus

$$\psi(\partial_{x_i}(h \otimes u)) = \sigma(\partial_{x_i})\psi(h \otimes u),$$

as required. One may similarly check that the action of ∂_{y_j} is compatible with ψ .

The next corollary is a combination of Theorem 3.1 and Corollary 9.2.4.

3.3 COROLLARY. *Let M be a finitely generated left A_{m+n}-module. Then G^*M and M have the same dimension.*

We will put all this together in a theorem.

THEOREM 3.4. *Let M be a left A_{m+n}-module. Then*

$$j^*M \cong M_\sigma/(Y)M_\sigma$$

as A_n-modules.

PROOF: As we have seen, $j^*M \cong \iota^*G^*M$. By Theorem 3.1, $G^*M \cong M_\sigma$. By §1,

$$\iota^*M_\sigma \cong M_\sigma/(Y)M_\sigma,$$

as claimed.

4. EXERCISES

4.1 Let $\iota : K \to K^2$ be the standard embedding. Compute the inverse image under ι of the following A_2-modules.

    (1) $A_2/A_2\partial_2$

    (2) $K[x_1, x_2]$

    (3) A_2/A_2x_2

    (4) $A_2/A_2x_2\partial_2$

    (5) $A_2/A_2\partial_2^3$

4.2 Which of the inverse images of Exercise 4.1 are finitely generated over A_1?

4.3 Let $\iota : X \to X \times K$ be the standard embedding, and denote by y the coordinate of K. Let A_{n+1} be the Weyl algebra generated by x's and ∂_x's and by y and ∂_y. Suppose that the term of highest *order* of $d \in A_{n+1}$ is ∂_y^k. Show that $\iota^*(A_{n+1}/A_{n+1}d)$ is a free module of rank k over A_n.

4.4 Keep the notations of Exercise 4.3. Suppose that I is a left ideal of A_{n+1} which properly contains $A_{n+1}d$. Show that $\iota^*(A_{n+1}/I)$ is finitely generated over A_n.

4.5 Let $F : X \to X \times X$ be the polynomial map defined by $F(X) = (X, X)$. This is a generalized embedding in the sense of §3. Let M and N be left modules over A_n. Then $M \widehat{\otimes} N$ is a left module over $A_{2n} = A_n \widehat{\otimes} A_n$. Show that

$$F^*(M \widehat{\otimes} M) \cong M \otimes_{K[X]} N.$$

The latter was defined in Exercise 14.4.4.

4.6 Show that if $\iota : X \to X \times Y$ is the standard embedding, then

$$\iota^*(A_{m+n}) \cong K[\partial_y] \widehat{\otimes} A_n$$

where $K[\partial_y]$ is the left A_m-module $K[\partial_{y_1}, \ldots, \partial_{y_m}]$.

CHAPTER 16

DIRECT IMAGES

In the previous chapter we saw that starting with an A_m-module and a polynomial map $F : K^n \to K^m$ we can construct an A_n-module, its inverse image. We shall now study a similar construction, which uses F to associate an A_m-module to an A_n-module, its direct image. Curiously, the direct image is easier to define for right modules; and it is with them that we start. Throughout this chapter, the conventions of 14.1.2 remain in force.

1. RIGHT MODULES

Let us briefly recall the definition of the inverse image. Let $F : X \to Y$ be a polynomial map. Let M be a left A_m-module. The inverse image of M under F is $F^*M = K[X] \otimes_{K[Y]} M$. This is a $K[X]$-module. It becomes an A_n-module with the ∂_{x_i} acting according to the formula

$$\partial_{x_i}(h \otimes u) = \frac{\partial h}{\partial x_i} \otimes u + \sum_1^m h \frac{\partial F_j}{\partial x_i} \otimes \partial_{y_j} u.$$

Let us rewrite this definition in a slightly different way. Since $A_m \otimes_{A_m} M \cong M$, we have that

$$F^*M \cong K[X] \otimes_{K[Y]} A_m \otimes_{A_m} M.$$

Writing $D_{X \to Y}$ for $F^*A_m = K[X] \otimes_{K[Y]} A_m$, one has that

$$F^*M = D_{X \to Y} \otimes_{A_m} M.$$

Note that $D_{X \to Y}$ is a left A_n-module and a right A_m-module. One easily checks, that these two structures are compatible and that it is in fact an A_n-A_m-bimodule.

The bimodule $D_{X \to Y}$ is the key to the direct image construction. Let N be a right A_n-module. The tensor product

$$F_*N = N \otimes_{A_n} D_{X \to Y}$$

is a right A_m-module, which is called the *direct image* of N under the polynomial map F.

Consider the projection $\pi : X \times Y \to Y$ defined by $\pi(X, Y) = Y$. From Ch.14, §3, it follows that

$$\pi^* A_m \cong K[X] \widehat{\otimes} A_m.$$

As a left A_n-module, $K[X]$ is isomorphic to $A_n / \sum_1^n A_n \partial_{x_i}$. Hence, by Corollary 13.2.3,

$$\pi^* A_m \cong A_{m+n} / \sum_1^n A_{m+n} \partial_{x_i}$$

is an isomorphism of A_{m+n}-A_m-bimodules. Now if N is a right A_{m+n}-module, then by Corollary 12.4.7,

$$\pi_* N \cong N / \sum_1^n N \partial_{x_i}$$

as right A_m-modules.

2. Transposition

We will now see how one can turn right modules into left modules, and vice versa. That will allow us to do direct images for left modules. We present the construction for general algebras.

Let R be a K-algebra. A *transposition* of R is an isomorphism $\tau : R \to R$ of the underlying K-vector space that satisfies

(1) $\tau(ab) = \tau(b)\tau(a)$,
(2) $\tau^2 = \tau$.

Examples of transpositions abound. The most familiar occurs in the matrix ring over a field. An example in the Weyl algebra is provided by the map $\tau : A_n \to A_n$ defined by

$$\tau(h \partial^\alpha) = (-1)^{|\alpha|} \partial^\alpha h,$$

where $h \in K[x_1, \ldots, x_n]$ and $\alpha \in \mathbb{N}^n$. We will refer to it as the *standard transposition* of A_n. If $d \in A_n$, we also use the notation d^τ for $\tau(d)$.

Let us go back to the general construction. Let R be a K-algebra and τ a transposition of R. If N is a right R-module then we define a left R-module N^t as follows. As an abelian group, $N^t = N$. If $a \in R$ and $u \in N^t$ then the left action of a on u is defined by $a \diamond u = u\tau(a)$. The left action $\diamond$ is called the *transposed action*. Instead of checking in detail that N^t is a left R-module, let us prove just one typical property, namely

$$(ab) \diamond u = a \diamond (b \diamond u)$$

where $a, b \in R$ and $u \in N$. The left hand side is, by definition, $u\tau(ab)$. Since τ is a transposition, it equals $u\tau(b)\tau(a)$. But this is $a \diamond (b \diamond u)$.

The same construction can be used to turn left modules into right modules. Indeed, let M be a left R-module. The right R-module M^t equals M as an abelian group. The right action of $a \in R$ on $u \in M$ is defined by $u \diamond a = \tau(a)u$. Note that since $\tau^2 = \tau$, the two constructions above are inverse to each other. If N is a right R-module, then $(N^t)^t = N$; and the same holds for left modules.

It is very easy to transpose the action when the module is cyclic. Let R be a K-algebra with a transposition τ . If J is a left ideal of R, define $J^t = \{\tau(a) : a \in J\}$. This is a right ideal of R.

2.1 PROPOSITION. *Let R be a K-algebra with a transposition τ and J a left ideal of R. The right R-module $(R/J)^t$ is isomorphic to R/J^t.*

PROOF: If $a \in R$, denote by $\overline{a}$ its image in $(R/J)^t$. Consider the map

$$\phi : R \longrightarrow (R/J)^t$$

defined by $\phi(a) = \overline{\tau(a)}$. It is clearly a homomorphism of additive groups. If $a, b \in R$, then

$$\phi(ab) = \overline{\tau(ab)},$$

which equals $\overline{\tau(b)\tau(a)} = \phi(a) \diamond b$. Thus ϕ is a homomorphism of right R-modules. One can easily check that it is surjective; let us calculate its kernel. Suppose that $a \in R$ satisfies $\phi(a) = 0$. Then $\tau(a) \in J$ or, equivalently, $a \in J^t$. Hence $\ker \phi \subseteq J^t$. The other inclusion is similarly proved, and we conclude that ker $\phi = J^t$. By the first homomorphism theorem, we have that $R/J^t \cong (R/J)^t$.

Let us illustrate the proposition with a concrete example of a module over the Weyl algebra. Let $J = A_n x_n$, and let τ be the standard transposition of A_n. Since $\tau(x_n) = x_n$, it follows that $J^t = x_n A_n$. Thus

$$(A_n/J)^t \cong A_n/x_n A_n.$$

This example will surface again later on.

All this can be extend to bimodules. Let R_1, R_2 be K-algebras with transpositions τ_1, τ_2, respectively. Let M be an R_1-R_2-bimodule. The *transposed* bimodule M^t is obtained by applying the above constructions to both the left and the right actions on M. Thus M^t and M have the same underlying abelian groups and, if $a_1 \in R_1$, $a_2 \in R_2$ and $u \in M$, then

$$a_2 \diamond u \diamond a_1 = (a_1)^{\tau_1} u (a_2)^{\tau_2}.$$

These actions are compatible with each other, since the original actions were so. Thus M^t is an R_2-R_1-bimodule.

The transposition of bimodules over the Weyl algebra has its own peculiarities. Suppose that $m \le n$ and that τ_m and τ_n are the standard transpositions of A_m and A_n, respectively. Then $A_m \subseteq A_n$ and τ_n restricted to A_m equals τ_m. This considerably simplifies the calculations. For instance, if J is a left ideal of A_n which is a right A_m-module, then it follows from these considerations and Proposition 2.1 that $(A_n/J)^t$ is isomorphic to A_n/J^t as an A_m-A_n-bimodule . The left ideal $J = A_n x_n$ is an example of this situation. This also means that we may denote the standard transposition simply by τ, irrespective of the index of the Weyl algebra in question.

For future reference, we must consider the behaviour of the transposition under tensor product.

2.2 LEMMA. *Let R_1, R_2 and R_3 be K-algebras with transposition. Suppose that M_1 is an R_1-R_2-bimodule and M_2 an R_2-R_3-bimodule. Then*

$$(M_1 \otimes_{R_2} M_2)^t \cong M_2^t \otimes_{R_2} M_1^t$$

as R_1-R_3-bimodules.

PROOF: Consider the map

$$\phi : M_2^t \times M_1^t \to (M_1 \otimes_{R_2} M_2)^t,$$

defined by $\phi(u_2, u_1) = u_1 \otimes u_2$, where $u_i \in M_i$. We show that ϕ is balanced, and leave the proof that it is bilinear to the reader. If $a \in R_2$ and τ_2 is the transposition of R_2, then

$$\phi(u_2 \diamond a, u_1) = \phi(\tau(a)u_2, u_1).$$

By the definition of ϕ,

$$\phi(u_2 \diamond a, u_1) = u_1 \otimes \tau(a)u_2.$$

Now, using that the tensor product $M_1 \otimes_{R_2} M_2$ is balanced, we get that

$$\phi(u_2 \diamond a, u_1) = u_1 \tau(a) \otimes u_2.$$

Hence,

$$\phi(u_2 \diamond a, u_1) = \phi(u_2, u_1 \tau(a)) = \phi(u_2, a \diamond u_1).$$

Thus ϕ induces a homomorphism of R_1-R_3-bimodules,

$$\overline{\phi} : (M_1 \otimes_{R_2} M_2)^t \to M_2^t \otimes_{R_2} M_1^t.$$

It is clear from the definition of $\overline{\phi}$ that it is bijective, and the proof is complete.

3. LEFT MODULES

Using the results of §2, we will do direct images for left modules. Let $F : X \to Y$ be a polynomial map. Consider the A_n-A_m-bimodule $D_{X \to Y}$. Using the standard transposition for A_m and A_n defined in the previous section, put

$$D_{Y \leftarrow X} = \left(D_{X \to Y} \right)^t.$$

This is an A_m-A_n-bimodule.

Let M be a left A_n-module. The *direct image* of M by F is defined by the formula

$$F_* M = D_{Y \leftarrow X} \otimes_{A_n} M.$$

It is clearly an A_m-module. This is very convenient, because to compute direct images one has only to transpose the actions of $D_{X \rightarrow Y}$; and this is usually easy to do.

Let us work out the calculations for the projection $\pi : X \times Y \rightarrow Y$. From §1, we have that

$$D_{X \times Y \rightarrow Y} \cong A_{m+n} / \sum_1^n A_{m+n} \partial_{x_i}.$$

Since the standard transposition of A_{m+n} maps ∂_{x_i} to $-\partial_{x_i}$, we conclude by Proposition 2.1 that

$$D_{Y \leftarrow X \times Y} \cong A_{m+n} / \sum_1^n \partial_{x_i} A_{m+n}.$$

Note that the y's commute with the ∂_x's. In particular, $\sum_1^n \partial_{x_i} A_{m+n}$ is a left A_m- submodule of A_{m+n}. Thus $A_{m+n} / \sum_1^n \partial_{x_i} A_{m+n}$ is a left A_m-module for the quotient action.

The direct image under projections does *not* take finitely generated modules to finitely generated modules. Indeed, A_{m+n} is cyclic as a module over itself, but $A_{m+n} / \sum_1^n \partial_{x_i} A_{m+n}$ is not finitely generated over A_m. It is a free A_m-module; with a basis given by the monomials on the x's.

Using this description of $D_{Y \leftarrow X \times Y}$ we can easily calculate the direct image of any module under a projection. If M is a left A_{m+n}-module, then by Corollary 12.4.7,

$$\pi^* M \cong M / \sum_1^n \partial_{x_i} M.$$

Another interesting example is provided by polynomial isomorphisms. Let $F : X \rightarrow Y$ be a polynomial map, and consider the polynomial isomorphism $G : X \times Y \rightarrow X \times Y$ defined by

$$G(X, Y) = (X, Y + F(X)).$$

In Ch. 15, §3, we saw how to calculate inverse images under this map. In particular, there exists an automorphism σ of A_{m+n} such that

$$D_{X \times Y \to X \times Y} = G^*(A_{m+n}) \cong (A_{m+n})_\sigma.$$

We wish to transpose this module in order to calculate direct images. The final result is very simple, but a technical lemma is required.

3.1 LEMMA. *Let τ be the standard transposition of A_{m+n}. Then $\tau \cdot \sigma \cdot \tau = \sigma$.*

PROOF: It is enough to check that the identity holds on the generators of A_{m+n}. First, τ is the identity on $K[X, Y]$ and σ restricts to an automorphism of $K[X, Y]$. Thus $\tau \cdot \sigma \cdot \tau = \sigma$ in $K[X, Y]$. On the other hand,

$$\tau \sigma \tau(\partial_{x_i}) = \tau \sigma(-\partial_{x_i}) = -\tau(\partial_{x_i} + \sum_1^m \frac{\partial F_j}{\partial x_i} \partial_{y_j}).$$

Since $\partial F_j / \partial x_i$ commutes with ∂_{y_j} we have that

$$\tau \sigma \tau(\partial_{x_i}) = \sigma(\partial_{x_i}).$$

Similarly, $\tau \sigma \tau(\partial_{y_j}) = \sigma(\partial_{y_j})$; and the proof is complete.

3.2 PROPOSITION. *Let M be a left A_{m+n}-module, then*

$$G_* M \cong M_{\sigma^{-1}}.$$

PROOF: Let us first find the transpose of $(A_{m+n})_\sigma$. Consider the map

$$\psi : ((A_{m+n})_\sigma)^t \to A_{m+n}$$

defined by $\psi(u) = \tau(u)$. It is a K-vector space isomorphism. Furthermore, if $a, b \in A_{m+n}$ and $u \in ((A_{m+n})_\sigma)^t$ then

$$a \diamond u \diamond b = \tau(b) \star u\tau(a) = \sigma(\tau(b))u\tau(a).$$

Therefore, $\psi(a \diamond u \diamond b) = a\tau(u)\tau\sigma\tau(b)$. By Lemma 3.1, $\tau \sigma \tau(b) = \sigma(b)$; and so

$$\psi(a \diamond u \diamond b) = a\tau(u)\sigma(b) = a\psi(u) \star b.$$

Hence $D_{X \times Y \leftarrow X \times Y}$ is A_{m+n} as a bimodule over itself, but with the right action twisted by σ. Therefore,

$$G_* M \cong A_{m+n} \otimes_\sigma M.$$

By Lemma 14.1.1 this is isomorphic to $M_{\sigma^{-1}}$.

3.3 COROLLARY. *Let M be a finitely generated left A_{m+n}-module. Then G_*M and M have the same dimension.*

PROOF: Follows from Proposition 3.2 and Corollary 9.2.4.

4. EXERCISES

4.1 Let R be a K-algebra and τ_1, τ_2 two transpositions of R. Show that $\tau_1(\tau_2)^{-1}$ is an automorphism of R.

4.2 Let $\pi : K^2 \to K$ be the projection on the second coordinate. Compute the direct image of the following A_2-modules under π.

(1) $A_2/A_2\partial_2$
(2) $K[x_1, x_2]$
(3) $A_2/A_2\partial_1$
(4) A_2/A_2x_2
(5) $A_2/A_2\partial_2^3$

4.3 Which of the direct images of Exercise 4.2 are finitely generated over A_1?

4.4 Let $F : X \to Y$ be a polynomial map and M a left A_n-module. Show that

$$F_*M \cong \left(M^t \otimes_{A_m} D_{X \to Y} \right)^t.$$

4.5 Let $\iota : X \to X \times Y$ be the standard embedding and $\eta : X \times Y \to X$ be the projection on the *first* coordinate. Let M be a left A_{m+n}-module. Show that ι^*M is the Fourier Transform of η_*M.

4.6 Let $F : X \to Y$ be a polynomial map. Define $G : X \times Y \to X \times Y$ by $G(X, Y) = (X, Y + F(X))$. Let M be a left A_{m+n}-module. Show that

$$G_*G^*M = M = G^*G_*M.$$

CHAPTER 17

KASHIWARA'S THEOREM

In this chapter we deal with the direct image under an embedding. This will lead us to an important structure theorem for a whole class of A_n-modules. The noncommutativity of the Weyl algebra will be an essential ingredient in these results. In fact the theorem fails to hold for polynomial rings, as we show at the end of §2. Throughout this chapter, the notation of 14.1.2 will be in force.

1. EMBEDDINGS

Let $\iota: X \rightarrow X \times Y$ be the standard embedding: $\iota(X) = (X, 0)$, where 0 is the origin of Y. It follows from Ch. 15, §1, that

$$D_{X \rightarrow X \times Y} = \iota^*(A_{m+n}) \cong A_{m+n}/(Y)A_{m+n},$$

where (Y) is the ideal of $K[X, Y]$ generated by the y's. One easily verifies that this isomorphism preserves the right A_{m+n}-module structure of $\iota^*(A_{m+n})$. Transposing this bimodule with the help of the standard transposition of A_{m+n} and using Proposition 16.2.1,

$$D_{X \times Y \leftarrow X} \cong A_{m+n}/A_{m+n}(Y).$$

This is an $A_{m+n} - A_n$-bimodule. If N is a left A_n-module, then its direct image under ι is, by definition, the module

$$\iota_* N = D_{X \times Y \leftarrow X} \otimes_{A_n} N.$$

Let us consider the structure of $D_{X \times Y \leftarrow X}$ in greater detail. By Corollary 13.2.3 there exists an isomorphism of bimodules,

$$D_{X \times Y \leftarrow X} \cong (A_m/A_m(Y)) \widehat{\otimes} A_n.$$

But $A_m/A_m(Y)$ is isomorphic to $K[\partial_{y_1}, \ldots, \partial_{y_n}] = K[\partial_y]$ as an A_m-module. Therefore,

(1.1) $$D_{X \times Y \leftarrow X} \cong K[\partial_y] \widehat{\otimes} A_n.$$

Let $\alpha \in \mathbb{N}^n$. Recall that the action of y_j on $\partial^\alpha \in K[\partial_y]$ is given by

$$y_j \cdot \partial^\alpha = -\alpha_j \partial^{\alpha - e_j}.$$

It is worth stressing that on the right hand side of (1.1), the y's and ∂_y's act only on $K[\partial_y]$, whilst the x's and ∂_x's act only on A_n; as in Ch. 13, §2. Thus if N is a left A_n-module,

$$\iota_* N \cong K[\partial_y] \widehat{\otimes} N.$$

It is very easy to calculate with the direct image in this form; we give two examples. The next theorem should be compared with Theorem 14.3.1.

1.2 THEOREM. *Let* $\iota : X \rightarrow X \times Y$ *be the standard embedding. If M is a finitely generated left A_n-module, then:*

(1) $\iota_* M$ *is a finitely generated left A_{m+n}-module.*

(2) $d(\iota_* M) = m + d(M).$

(3) $m(\iota_* M) \leq m(M).$

PROOF: (1) follows from Proposition 13.2.1. Since $K[\partial_y]$ is the Fourier transform of $K[X]$, it must have dimension m and multiplicity 1 by Proposition 9.2.2. Thus (2) and (3) follow from Theorem 13.4.1.

The following corollary is an immediate consequence of the theorem.

1.3 COROLLARY. *Let M be a holonomic A_n-module and ι the standard embedding. Then $\iota_* M$ is a holonomic A_{m+n}-module.*

We now turn to another application.

1.4 LEMMA. *Let N be a left A_n-module. Every element of $\iota_* N$ is annihilated by a power of y_j, where $1 \leq j \leq m$.*

PROOF: Without loss of generality assume that $j = 1$. Since

$$y_1 \cdot \partial^\alpha = -\alpha_1 \partial^{\alpha - e_1}$$

in $K[\partial_y]$, it follows that ∂^α is annihilated by y_1^k, for any $k \geq \alpha_1 + 1$. Now every element of $\iota_* N$ may be written in the form

$$\sum_{\alpha \in I} (\partial^\alpha \otimes u_\alpha),$$

where $u_\alpha \in N$ and $I \subseteq \mathbb{N}^n$. Choose an integer k such that $k \geq \alpha_1 + 1$, for every $\alpha \in I$. Then $y_1^k \cdot \partial^\alpha = 0$, for every $\alpha \in I$. Thus

$$y_1^k \cdot \left(\sum_{\alpha \in I} \partial^\alpha \otimes u_\alpha \right) = 0.$$

We also deduce from the isomorphism

$$D_{X \times Y \leftarrow X} \cong K[\partial_y] \widehat{\otimes} A_n$$

that $D_{X \times Y \leftarrow X}$ is free of infinite rank as a right A_n-module. This follows from the fact that $K[\partial_y]$ is a K-vector space of infinite dimension. The monomials in the ∂_y's form a basis for $D_{X \times Y \leftarrow X}$ as a right A_n-module. A simple consequence of this fact and Theorem 12.4.2 is the following corollary.

1.5 COROLLARY. *If N is a left A_n-module, then $\iota_* N = 0$ if and only if $N = 0$.*

2. KASHIWARA'S THEOREM

In this section we study the direct image under the following special case of the standard embedding $\iota : X \to X \times K$. Let y denote the coordinate of K. The polynomial ring in the x's and y will be denoted by $K[X, y]$. Let M be a $K[X, y]$-module and H the hyperplane of equation $y = 0$ in $X \times K$. The submodule of M of elements with *support* on H is defined by

$$\Gamma_H M = \{ u \in M : u \text{ is annihilated by a power of } y \}.$$

Note that by Proposition 12.5.1, this is exactly the kernel of the map $M \to M[y^{-1}]$. Although we have defined $\Gamma_H M$ for $K[X, y]$-modules, in most of our applications it will also be an A_{n+1}-module.

2.1 PROPOSITION. *Let M be an A_{n+1}-module. Then $\Gamma_H M$ is a submodule of M.*

PROOF: Let $u_1, u_2 \in \Gamma_H M$. Suppose that y^{k_1} and y^{k_2} are the powers that annihilate u_1, u_2, respectively. Then $y^{k_1 + k_2}$ annihilates both u_1 and u_2, hence their sum.

It is harder to show that $\Gamma_H M$ is closed under multiplication by elements of A_{n+1}. Of course it is enough to check that it is closed for multiplication by the generators of A_{n+1}. Assume that $u \in \Gamma_H M$ is annihilated by y^k. Note that y^k commutes with the x's, ∂_x's and with y. Thus,

$$y^k(\partial_{x_i} u) = \partial_{x_i}(y^k u) = 0.$$

and so $\partial_{x_i} u \in \Gamma_H M$. Similarly, $x_i u, \, yu \in \Gamma_H M$. Hence we need only check that $\partial_y u \in \Gamma_H M$. Using the Weyl algebra relations,

$$y^{k+1}(\partial_y u) = \partial_y y^{k+1} u - (k+1)y^k u = 0,$$

which completes the proof of the proposition.

Given an A_{n+1}-module M, put

$$\ker_M(y) = \{u \in M : yu = 0\}.$$

Clearly $\ker_M(y) \subseteq \Gamma_H M$. Although $\ker_M(y)$ is closed under addition, it is not an A_{n+1}- submodule of M. But, since the y's commute with the x's and ∂_x's, it is an A_n-submodule of M. We now make explicit the true relation between $\ker_M y$ and $\Gamma_H M$. It all begins with a lemma. To simplify the notation, let $M_0 = \ker_M(y)$.

2.2 LEMMA. *Let M be a left A_{n+1}-module, and let H be the hyperplane $y = 0$. Then*

(1) *For each $k \in \mathbb{N}$, the map*

$$\partial_y : \partial_y^k M_0 \longrightarrow \partial_y^{k+1} M_0$$

defined by right multiplication by ∂_y is injective.

(2) $y(A_{n+1} M_0) = A_{n+1} M_0.$

(3) $A_{n+1} M_0 = M_0 \oplus \partial_y M_0 \oplus \partial_y^2 M_0 \oplus \cdots.$

PROOF: Let $u \in M_0$ and k be a positive integer. We have that $[y, \partial_y^k] = -k\partial_y^{k-1}$ and that y annihilates u, thus $y \cdot \partial_y^k u = -k\partial_y^{k-1} u$. Therefore,

(2.3) $$y^i \cdot \partial_y^k u = (-1)^i k \ldots (k - i + 1)\partial_y^{k-i} u$$

for every $i \geq 0$. Suppose that $\partial_y^k u = 0$. Applying (2.3) with $i = k$ to this equation, we conclude that $u = 0$. This proves (1).

Now, every element of A_{n+1} can be written in the form

$$\sum_0^k \partial_y^i b_i,$$

where $b_i \in A_n[y]$. Since M_0 is a left A_n-module that is annihilated by y, the elements of $A_{n+1} M_0$ can be written as

$$\sum_0^k \partial_y^i u_i$$

where $u_i \in M_0$. In other words,

$$A_{n+1} M_0 = M_0 + \partial_y M_0 + \partial_y^2 M_0 + \dots.$$

Thus (2) follows from $\partial_y^k u = -(k+1)^{-1} y \cdot (\partial_y^{k+1} u)$.

Assume now that

$$u_0 + \partial_y u_1 + \dots + \partial_y^k u_k = 0,$$

where $u_0, \dots, u_k \in M_0$. Multiplying by y^k and using (2.3) we conclude that $u_k = 0$. Thus we have a sum with $k - 1$ terms. By induction on the number of terms, we conclude that $u_0 = \dots = u_k = 0$. Hence the sum is direct, and (3) is proved.

We are now ready to state a preliminary version of Kashiwara's theorem.

2.4 THEOREM. *Let M be a left A_{n+1}-module, let H be the hyperplane $y = 0$ and denote by $\iota : X \to X \times K$ the standard embedding. The A_{n+1}-modules $\iota_*(\ker_M y)$ and $\Gamma_H M$ are isomorphic.*

PROOF: We have seen that $M_0 = \ker_M y$ is an A_n-submodule of M. It follows from §1 that

$$\iota_*(M_0) = K[\partial_y] \widehat{\otimes} M_0.$$

We wish to show that this is isomorphic to $\Gamma_H M$.

Since $M_0 \subseteq \Gamma_H M$, we can use the universal property of the tensor product to define a map

$$\phi : K[\partial_y] \widehat{\otimes} M_0 \to \Gamma_H M$$

by $\phi(f \otimes u) = fu$, where f is a polynomial in ∂_y. The image of ϕ is $A_{n+1} M_0$ and its kernel is zero, by Lemma 2.2(3). Thus to prove that ϕ is surjective it is enough to show that $\Gamma_H M$ is contained in the A_{n+1}-submodule of M generated by M_0.

Let $u \in \Gamma_H M$ and assume that $y^k u = 0$. We want to show that $u \in A_{n+1} M_0$. The proof is by induction on k. If $k = 1$, then $u \in M_0$. Suppose that the result holds for every element of M_0 annihilated by y^{k-1}. From $y^k u = 0$, we deduce that $\partial_y(y^k u) = 0$. Since $[\partial_y, y^k] = k y^{k-1}$,

$$y^{k-1}(ku + y\partial_y u) = 0.$$

Thus, by the induction hypothesis, $ku + y\partial_y u \in A_{n+1} M_0$. But $y^{k-1}(yu) = y^k u = 0$ and so $yu \in A_{n+1} M_0$. Since $A_{n+1} M_0$ is an A_{n+1}-submodule of M, we have that

$$ku + y\partial_y u - \partial_y y u \in A_{n+1} M_0.$$

But this expression is equal to $(k-1)u$, because $[\partial_y, y] = 1$. Thus $u \in A_{n+1} M_0$, as required.

Combining the fact that $\Gamma_H(M) = A_{n+1} M_0$ with Lemma 2.2(2) we get the following result, which will be used in the next chapter.

2.5 COROLLARY. *Let M be a left A_{n+1}-module and let H be the hyperplane $y = 0$. Then $y\Gamma_H(M) = \Gamma_H(M)$.*

We say that a $K[X, y]$-module M has *support* on H if $\Gamma_H M = M$. In particular this definition may be applied to A_{n+1}-modules. The next result is a structure theorem for A_{n+1}-modules with support on H. The proof is an immediate consequence of Theorem 2.4 and the above definition.

2.6 COROLLARY. *Let M be an A_{n+1}-module with support on H. Then $M \cong \iota_*(\ker_M y)$.*

These results hold in greater generality. In Ch. 18, §3 we will use the language of category theory to bring forth the true nature of Kashiwara's theorem.

We have taken care to point out that both $\Gamma_H M$ and $\ker_M y$ are defined for $K[X, y]$-modules as well as for A_{n+1}-modules. In fact there is also a natural way to define direct images for modules over polynomial rings under polynomial maps; see Exercise 3.4. Applied to the standard embedding ι : $X \to X \times K$, the direct image $\iota_+ M$ of a $K[X]$-module M equals M, as a vector space. The action of $f(X, Y) \in K[X, y]$ on $u \in M$ is given by $fu = f(X, 0)u$.

For this direct image, it is *not* true that if M is a $K[X, y]$-module then

$$\iota_+(\ker_M y) \cong \Gamma_H M.$$

Here is a simple example. Consider the $K[X, y]$-module $M = K[X, y]/(y^2)$. Since every element of M is annihilated by y^2, it follows that $\Gamma_H M = M$. But a straightforward calculation shows that $\ker_M y = yM$. Since $\iota_+(\ker_M y) = yM$, as vector spaces, we have that $\iota_+(\ker_M y) \neq \Gamma_H M$.

3. EXERCISES

3.1 Let $\iota : K \to K^2$ be the standard embedding given by $\iota(x) = (x, 0)$. Compute the direct image of the following modules under ι.

(1) $K[x]$
(2) $A_1/A_1\partial^5$
(3) $A_1/A_1 x\partial$
(4) $A_1/A_1(x^2\partial + 4x + 1)$

3.2 Let $\iota : X \to X \times Y$ be the standard embedding. Let M be a left A_n-module. Is it true that

$$\iota_*(M_{\mathcal{F}}) \cong (\iota_* M)_{\mathcal{F}}?$$

3.3 Let $\iota : X \to X \times K$ be the standard embedding and let M be a left A_{n+1}-module with support on the hyperplane H of equation $y = 0$. Show that $\ker_M y$ is isomorphic, as a left A_n-module, to

$$Hom_{A_{n+1}}(D_{X \times K \leftarrow X}, M).$$

3.4 The purpose of this exercise is to define a direct image for modules over polynomial rings. Suppose that $F : X \to Y$ is a polynomial map. The comorphism of F is a homomorphism of rings $F^\sharp : K[Y] \to K[X]$. Let M be a $K[X]$-module. We want to construct a $K[Y]$-module F_+M. The definition is as follows. As K-vector spaces, $F_+M = M$. The action of $g \in K[Y]$ on $u \in M$ is defined by $gu = F^\sharp(g)u$.

(1) Check that this action makes F_+M into a $K[Y]$-module.

(2) Compute ι_+M for the standard embedding $\iota : X \to X \times Y$.

(3) Let N be a $K[X, Y]$-module. Compute π_+N for the projection $\pi : X \times Y \to Y$ onto the second coordinate.

(4) If M is finitely generated over $K[X]$, is F_+M necessarily finitely generated over $K[Y]$?

3.5 Let M be a left A_n-module and $\phi : M \to M[x_n^{-1}]$ the natural embedding. Show that $\ker \phi$ and $\operatorname{coker}\phi$ are supported on the hyperplane $x_n = 0$.

3.6 Let M be an irreducible A_n-module. For a non-zero element u in M let $J(u) = \{f \in K[X] : fu = 0\}$. Show that:

(1) If $0 \neq v \in M$ then $radJ(v) = radJ(u)$.

(2) $radJ(u)$ is a prime ideal of $K[X]$.

(3) If M is supported on the hypersurface $x_n = 0$, then $radJ(u)$ is contained in the ideal of $K[X]$ generated by x_n.

CHAPTER 18

PRESERVATION OF HOLONOMY

We have seen in the previous chapters that holonomic modules are preserved by inverse images under projections and by direct images under embeddings. However, as we also saw, inverse images under embeddings and direct images under projections do *not* preserve the fact that a module is finitely generated. Fortunately, though, holonomic modules are preserved by all kinds of inverse and direct images. The proof of this result will use all the machinery that we have developed so far. It gives yet one more way to construct examples of holonomic modules. We retain the notations of 14.1.2.

1. INVERSE IMAGES

The key to the results in this chapter is a decomposition of polynomial maps in terms of embeddings and projections. The idea goes back to A. Grothendieck.

Let $F : X \to Y$ be a polynomial map. We may decompose F as a composition of three polynomial maps: a projection, an embedding and an isomorphism. The maps are the following. The projection is $\pi : X \times Y \to Y$, defined by $\pi(X, Y) = Y$. The isomorphism is $G : X \times Y \to X \times Y$ where $G(X, Y) = (X, Y + F(X))$. Finally, the embedding is $\iota : X \to X \times Y$, defined by $\iota(X) = (X, 0)$. One can immediately check that $F = \pi \cdot G \cdot \iota$.

Let us consider the effect of an inverse image under F on the holonomic A_m-module M. By Theorem 15.2.1,

$$F^* M \cong \iota^*(G^*(\pi^* M)).$$

It follows from Corollary 14.3.2 and Corollary 15.3.3 that $G^*(\pi^* M)$ is holonomic if M is holonomic. Hence $F^* M$ will be a holonomic module if we can prove that ι^* preserves holonomy. We shall now do this.

We will begin with the standard embedding $\iota : X \to X \times K$. The coordinates of $X \times K$ will be denoted by $x_1, \ldots, x_n, y$ and H will stand for the hyperplane $y = 0$.

1.1 LEMMA. *Let M be a left A_{n+1}-module. Put $M' = M/\Gamma_H M$. Then*

$$\iota^* M \cong \iota^* M'.$$

PROOF: The inverse image $\iota^* M$ is isomorphic to M/yM. Similarly $\iota^* M' \cong M'/yM'$. Since M' is a quotient of M, there exists a surjective map of A_n-modules,

$$\phi : M/yM \longrightarrow M'/yM'.$$

Let $u \in M$ and suppose that $\phi(u + yM) = 0$. Then $u \in yM + \Gamma_H M$. Since $y\Gamma_H M = \Gamma_H M$ by Corollary 17.2.5; we conclude that $u \in yM$ and that ϕ is an isomorphism.

1.2 LEMMA. *Let M be a holonomic A_{n+1}-module which does not contain any non-zero element with support on H. Then $\iota^* M$ is a holonomic A_n-module whose multiplicity cannot exceed the multiplicity of M.*

PROOF: Let Γ be a good filtration for M. Put

$$\Omega_j = (\Gamma_j + yM)/yM.$$

Note that we are using a filtration of M – an A_{n+1}-module – to construct a filtration of M/yM – an A_n-module. It is clear that $\Omega_j \subseteq \Omega_{j+1}$ and that $\bigcup_{j \geq 0} \Omega_j = M/yM$. Since $B_i(A_n) \subseteq B_i(A_{n+1})$, we have that $B_i(A_n)\Gamma_j \subseteq \Gamma_{i+j}$. Thus

$$B_i(A_n)\Omega_j \subseteq \Omega_{i+j}.$$

Hence $\Omega = \{\Omega_j : j \geq 0\}$ is a filtration of the A_n-module M/yM. Note that we do not know whether this filtration is good.

By the third homomorphism theorem for vector spaces,

$$\Omega_j \cong \Gamma_j/(\Gamma_j \cap yM).$$

Therefore,

$$\dim_K \Omega_j = \dim_K \Gamma_j - \dim_K(\Gamma_j \cap yM).$$

But $y\Gamma_{j-1} \subseteq \Gamma_j \cap yM$, and so

$$\dim_K \Omega_j \leq \dim_K \Gamma_j - \dim_K(y\Gamma_{j-1}).$$

Since M does not have any element supported on H, the map $\Gamma_{j-1} \rightarrow y\Gamma_{j-1}$ given by right multiplication by y is injective. Hence $\dim_K(y\Gamma_{j-1}) = \dim_K(\Gamma_{j-1})$ and so

(1.3) $$\dim_K \Omega_j \leq \dim_K \Gamma_j - \dim_K(\Gamma_{j-1}).$$

Now let $\chi(t, M, \Gamma)$ be the Hilbert polynomial of M with respect to the filtration Γ. For $j \gg 0$, it follows from (1.3) that

$$\dim_K \Omega_j \leq \chi(j, M, \Gamma) - \chi(j-1, M, \Gamma).$$

The polynomials $\chi(t, M, \Gamma)$ and $\chi(t-1, M, \Gamma)$ have the same leading term,

$$m(M) t^n / n!,$$

where $m(M)$ stands for the multiplicity of M. Hence their difference has leading term $m(M) t^{n-1} / (n-1)!$. Thus there exists $c \in \mathbb{Q}$ such that

$$\dim_K \Omega_j \leq \frac{m(M) j^{n-1}}{(n-1)!} + c(j+1)^{n-2}.$$

We conclude from Lemma 10.3.1 that M/yM is a holonomic module whose multiplicity cannot exceed $m(M)$.

1.4 THEOREM. *Let* $\iota : X \rightarrow X \times Y$ *be the standard embedding and let* M *be a holonomic* A_{m+n}-*module. Then* $\iota^* M$ *is a holonomic* A_n-*module.*

PROOF: Recall that $Y = K^m$. We proceed by induction on m. Suppose first that $m = 1$. Denote by H the hyperplane $y_1 = 0$, and put $M' = M/\Gamma_H M$. The module M' is a quotient of a holonomic module, hence is holonomic. Since M' does not contain any element supported on H, it follows from Lemma 1.2 that $\iota^* M'$ is holonomic. But by Lemma 1.1, $\iota^* M \cong \iota^* M'$, hence $\iota^* M$ is also holonomic.

Now the standard embedding $\iota : X \rightarrow X \times Y$ may be written as a composition of two standard embeddings, namely $\iota_1 : X \rightarrow X \times K^{n-1}$ and $\iota_2 : X \times K^{n-1} \rightarrow X \times Y$. By Theorem 15.2.1,

$$\iota^* M = \iota_1^* \iota_2^* M.$$

The result follows by induction.

Let us now put all these results together.

1.5 THEOREM. *Let $F : X \longrightarrow Y$ be a polynomial map. If M is a holonomic A_m-module, then F^*M is a holonomic A_n-module.*

2. DIRECT IMAGES.

That holonomy is preserved under direct images is proved in a very similar way to the inverse image case. However we must first tidy up some loose ends.

2.1 THEOREM. *Let $F : X \longrightarrow Y$ and $G: Y \longrightarrow Z$ be two polynomial maps. Let M be a left A_n-module. Then*

$$(GF)_*M \cong G_*F_*M.$$

PROOF: It follows from Theorem 15.2.1 that $(GF)^*(A_r) \cong F^*G^*(A_r)$. In the notation of Ch. 16, §1, this is equivalent to

$$D_{X \to Z} \cong D_{X \to Y} \otimes_{A_m} D_{Y \to Z}.$$

Applying Lemma 16.2.2 to this isomorphism we get that

$$D_{Z \leftarrow Y} \otimes_{A_m} D_{Y \leftarrow X} \cong D_{Z \leftarrow X}.$$

The theorem is an immediate consequence of this formula.

Now, in the notation of §1, we have that the polynomial map $F : X \longrightarrow Y$ can be written as $F = \pi \cdot G \cdot \iota$. Let M be a holonomic A_m-module. By Theorem 2.1,

$$F_*M = \pi_*(G_*(\iota_*M)).$$

We already know, from Corollaries 16.3.3 and 17.1.3, that $G_*(\iota_*M)$ is holonomic. In order to show that F_*M is holonomic, it is enough to show that π_* preserves holonomy.

2.2 THEOREM. *Let $\pi : X \times Y \longrightarrow Y$ be the projection $\pi(X, Y) = Y$. Let M be a holonomic A_{m+n}-module. Then π_*M is a holonomic A_m-module.*

PROOF: We know from Ch. 16, §3, that

$$\pi_*M \cong M / \sum_{1}^{n} \partial_{x_i}M.$$

By Proposition 5.2.1 this is the Fourier transform of $M/\sum_1^n x_i M$. Therefore the modules $\pi_* M$ and $M/\sum_1^n x_i M$ have the same dimension by Proposition 9.2.2. However, $M/\sum_1^n x_i M$ is isomorphic to $\iota^* M$ where $\iota : Y \to X \times Y$ is the embedding $\iota(Y) = (0, Y)$. By Theorem 1.4, $\iota^* M$ is holonomic; hence so is $\pi_* M$.

Summing up, we have proved the following theorem.

2.3 THEOREM. *Let $F : X \to Y$ be a polynomial map. If M is a holonomic A_n-module, then $F_* M$ is a holonomic A_m-module.*

3. Categories and functors.

We now introduce direct and inverse images as functors and generalize our previous version of Kashiwara's theorem using categories. For that we assume that the reader is familiar with the language of category theory. All that is required here can be found in [Cohn 79].

We shall be dealing with three categories. The category of left A_n-modules and module homomorphisms will be denoted by $\mathcal{M}^n$, and the full subcategory of finitely generated left A_n-modules by $\mathcal{M}_f^n$. The full subcategory of $\mathcal{M}^n$ whose objects are holonomic left A_n-modules will be denoted by $\mathcal{H}^n$. Since holonomic modules are finitely generated, it follows that $\mathcal{H}^n$ is a full subcategory of $\mathcal{M}_f^n$. Note that these are all abelian categories.

Let $F : X \to Y$ be a polynomial map. The inverse image F^* is a functor,

$$F^* : \mathcal{M}^m \to \mathcal{M}^n.$$

If M is an object in $\mathcal{M}^m$ then

$$F^* M \cong D_{X \to Y} \otimes_{A_m} M.$$

Thus F^* is a right exact functor by [Cohn 79, §4.3, Proposition 3]. In general, this functor is not exact. For example, let $\iota : K \to K^2$ be the standard embedding: $\iota(x) = (x, 0)$. Recall that $\iota^*(A_2) = A_2/x_2 A_2$. Let $\phi : A_2 \to A_2$ be the map defined by $\phi(d) = dx_2$. Applying the functor ι^* we get a homomorphism of A_1-modules,

$$\iota^*(\phi) : A_2/x_2 A_2 \to A_2/x_2 A_2.$$

But $1 + x_2 A_2$ is mapped to zero by $\iota^*(\phi)$; hence it cannot be injective. Thus ι^* is not an exact functor.

An important special case occurs when the polynomial map is a projection. Define $\pi : X \times Y \to Y$ by $\pi(X, Y) = Y$. Then

$$D_{X \times Y \to Y} \cong A_{m+n} / \sum_{1}^{n} A_{m+n} \partial_{x_i}$$

is free as a right A_m-module: the monomials in the x's form a basis for this module. Since free modules are flat, we conclude that the functor

$$\pi^* : \mathcal{M}^m \to \mathcal{M}^{m+n}$$

is exact. We have also seen that π^* preserves noetherianness. Hence we may restrict it to a functor,

$$\pi^* : \mathcal{M}_f^m \to \mathcal{M}_f^{m+n}.$$

This last statement does not hold true for general polynomial maps: it is false for embeddings, for example; see Ch. 15, §1. On the other hand, if $F : X \to Y$ is any polynomial map, then by Theorem 1.5 the inverse image F^* restricts to a functor from $\mathcal{H}^m$ to $\mathcal{H}^n$. So the behaviour of the inverse image functor in the holonomic category is up to expectations.

For direct images we have a similar situation. Given a polynomial map $F : X \to Y$, the direct image F_* determines a functor,

$$F_* : \mathcal{M}^n \to \mathcal{M}^m.$$

Since F_* is also defined by a tensor product, the direct image is right exact. It is not in general exact. Although the direct image does not always preserve finitely generated modules, it preserves holonomic modules by Theorem 2.3.

Let us turn to direct images under embeddings. Denoting the coordinates of $X \times K$ by $x_1, \ldots, x_n, y$, let $\iota : X \to X \times K$ be the standard embedding $\iota(X) = (X, 0)$. Then

$$D_{X \times K \leftarrow X} \cong A_{n+1} / A_{n+1} y$$

is free as a right A_n-module: the monomials in ∂_y form a basis. Hence the functor

$$\iota_* : \mathcal{M}^n \to \mathcal{M}^{n+1}$$

is exact. Since $D_{X \times K \leftarrow X}$ is finitely generated as a left A_{n+1}-module, the functor ι_* restricts to a functor $\mathcal{M}_f^n \to \mathcal{M}_f^{n+1}$.

Now let H be the hyperplane defined by the equation $y = 0$. Let $\mathcal{M}^{n+1}(H)$ be the full subcategory of $\mathcal{M}^{n+1}$ of modules with support on H. Thus, M is an object in $\mathcal{M}^{n+1}(H)$ if and only if $\Gamma_H M = M$. In Lemma 17.1.4 we showed that if M is an object in $\mathcal{M}^n$, then $\iota_* M$ has support on H. This leads to a categorical version of Kashiwara's theorem.

3.1 THEOREM. *Let $\iota : X \to X \times K$ be the embedding $\iota(X) = (X, 0)$ and H the hyperplane of equation $y = 0$. The direct image functor,*

$$\iota_* : \mathcal{M}^n \to \mathcal{M}^{n+1}(H),$$

is an equivalence of categories.

PROOF: Consider the functor

$$\mathcal{K} : \mathcal{M}^{n+1} \to \mathcal{M}^n$$

defined on a left A_{n+1}-module M by $\mathcal{K}(M) = \ker_M H$. By Theorem 17.2.4 we have that

$$\iota_*(\mathcal{K}(M)) = \Gamma_H M.$$

If M has support on H, then $\iota_*(\mathcal{K}(M)) = M$.

On the other hand, let N be any left A_n-module. Then

$$\iota_* N \cong K[\partial_y] \widehat{\otimes} N$$

by Ch. 17, §1. Since the only elements of $K[\partial_y]$ that are annihilated by y are the constants, we conclude that

$$\ker_{\iota_* N} H = K \widehat{\otimes} N \cong N$$

as A_n-modules. Hence $\mathcal{K}(\iota_* N) \cong N$.

We leave it to the reader to check that the natural isomorphisms defined above have the expected behaviour on maps.

This may be generalized to embeddings in higher dimension. Let $\iota : X \to X \times Y$ be the embedding $\iota(X) = (X, 0)$. We will identify Y with the subspace of equations $y_1 = \cdots = y_m = 0$. Let M be an object in $\mathcal{M}^{m+n}$. Write H_i for the hyperplane $y_i = 0$. We say that M has *support* on Y if it has support on H_i, for every $i = 1, \ldots, m$. The full subcategory of A_{m+n}-modules with support on Y will be denoted by $\mathcal{M}^{m+n}(Y)$.

3.2 COROLLARY. *The functor*

$$\iota_* : \mathcal{M}^n \to \mathcal{M}^{m+n}(Y)$$

is an equivalence of categories.

PROOF: Recall that $Y = K^m$. We proceed by induction on m. The case $m = 1$ has been proved in the theorem. Let $W = K^{m-1}$. There are embeddings $\iota_1 : X \to X \times W$ and $\iota_2 : X \times W \to X \times Y$ such that $\iota = \iota_2 \cdot \iota_1$. It follows by induction that $(\iota_1)_*$ and $(\iota_2)_*$ are equivalencies of categories. Since, by Theorem 2.1,

$$\iota_* = (\iota_2)_*(\iota_1)_*$$

we have that ι_* is an equivalence of categories.

Since the direct image under embeddings preserves finitely generated modules, we may replace the categories in Corollary 3.2 by the corresponding full subcategories of finitely generated modules. A similar statement holds for the holonomic categories, see Exercise 4.5.

Kashiwara's theorem can be used to prove a structure theorem for A_n-modules with support on the origin.

3.3 COROLLARY. *If M is a finitely generated left A_n-module with support on the origin, then there exists an integer $r \geq 0$ such that*

$$M \cong (K[\partial_1, \dots, \partial_n])^r.$$

PROOF: Let $\iota : \{0\} \to X$ be the standard embedding: $\iota(0) = 0$. Note first that it follows from the definitions that

$$K[\partial_1, \dots, \partial_n] \cong \iota_*(K).$$

Hence $K[\partial_1, \dots, \partial_n]$ has support on every hyperplane H_i, for $1 \leq i \leq n$, by Lemma 17.1.4. Therefore it has support on $H_1 \cap \cdots \cap H_n = \{0\}$.

On the other hand, if M has support on the origin then $\mathcal{K}(M)$ is a finitely generated module over $A_0 = K$. Hence $\mathcal{K}(M) \cong K^r$, for some $r \geq 0$. But $\iota_*(\mathcal{K}(M)) = M$, whilst

$$\iota_*(K^r) \cong (K[\partial_1, \dots, \partial_n])^r.$$

Therefore, $M \cong (K[\partial_1, \ldots, \partial_n])^r$, as required.

4. EXERCISES

4.1 Let M, N be holonomic A_n-modules. Show that $M \otimes_{K[X]} N$ is a holonomic A_n-module.

Hint: Exercise 15.4.5.

4.2 Let $p \in K[X]$ be a non-zero polynomial and let M be a holonomic A_n-module. Show that $M[p^{-1}] = K[X, p^{-1}] \otimes_{K[X]} M$ is a holonomic A_n-module.

4.3 Give an example to show that the direct image functor is not exact under projections.

4.4 Let $n \leq k \leq 2n$ be positive integers. Use Kashiwara's theorem and induction to construct an A_n-module of dimension k.

4.5 As in §3, let us identify Y with the linear subspace of equations $y_1 = \cdots = y_m = 0$ in $X \times Y$. Let $\mathcal{H}^{m+n}(Y)$ be the category of holonomic left A_{m+n}-modules with support on Y. Show that this category is equivalent to $\mathcal{H}^n$.

CHAPTER 19
STABILITY OF DIFFERENTIAL EQUATIONS

In this chapter we investigate the global asymptotic stability of a system of ordinary differential equations on the plane. This is closely related to the Jacobian conjecture. Holonomic modules will make a special appearance in §2.

1. ASYMPTOTIC STABILITY

We begin with some basic facts about the stability of singular points of systems of differential equations. Let

$$G : \mathbb{R}^n \to \mathbb{R}^n$$

be a function of class C^r for some $r > 2$, and assume that $G(0) = 0$. Consider the differential equation

$$(1.1) \qquad\qquad \dot{X} = G(X).$$

By the uniqueness theorem [Arnold **81**, Ch. 2, §8.3], the solution ϕ with initial condition $\phi(0) = 0$ is $\phi = 0$. We are interested in the behaviour of solutions with neighbouring initial conditions. The singular point $X = 0$ of equation (1.1) is *asymptotically stable* if:

(1) Given $\epsilon > 0$, there exists $\delta > 0$ (depending only on ϵ and not on t) such that, for every P_0 with $|P_0| < \delta$, the solution ϕ of (1.1) with initial condition $\phi(0) = P_0$ can be extended to the whole half line $t > 0$ and satisfies $|\phi(t)| < \epsilon$ for every $t > 0$.

(2) There exists $\eta > 0$ such that $\lim_{t \to +\infty} \phi(t) = 0$ for all solutions ϕ of (1.1) which satisfy $\phi(0) < \eta$.

Condition (1) above means that if the solution is initially within a ball of radius δ around the origin then it will never leave a ball of radius ϵ. Asymptotic stability is easy to determine for linear systems.

THEOREM 1.2. *Let A be an $n \times n$ matrix with entries in $\mathbb{R}$. The origin is an asymptotically stable singular point of $\dot{X} = A \cdot X$ if and only if all the eigenvalues of A have negative real part.*

Lyapunov showed that this can be extended to give a criterion to determine whether 0 is asymptotically stable in terms of the linearized system $\dot{X} = JG(0) \cdot X$.

THEOREM 1.3. *If the real part of every eigenvalue of $JG(0)$ is negative, then 0 is an asymptotically stable point of (1.1).*

For a proof of this theorem see [Arnold **81**, Ch. 3, Theorem 23.3]. We shall say that 0 is *globally asymptotically stable* if η may be taken to be ∞ in (2) above. For a linear system, if 0 is asymptotically stable, then it is globally asymptotically stable. Markus and Yamabe conjectured in [Markus and Yamabe **60**] the following criterion for global stability.

1.4 CONJECTURE. *The origin is globally asymptotically stable for (1.1) if, for each $P \in \mathbb{R}^n$, the origin is an asymptotically stable point of the system $\dot{X} = JG(P) \cdot X$.*

It is shown in [Gutierrez **93**] that the conjecture is true if $n = 2$. [Barabanov **88**] gives a counter-example for $n \geq 4$. In this chapter we will study the case $n = 2$ when G is polynomial, which was settled in [Meisters and Olech **88**].

Let us return to the hypothesis in Conjecture 1.4. For systems on the plane, the Jacobian $JG(P)$ is a 2×2 matrix. By Theorem 1.2 the system $\dot{X} = JG(P) \cdot X$ has the origin as an asymptotically stable point if and only if the eigenvalues of $JG(P)$ have negative real part. Note that since we are in the 2×2 case, this is equivalent to saying that the matrix $JG(P)$ has positive determinant and negative trace. This suggests a definition.

Let $\mathcal{F}$ be the class of C^1 maps $F : \mathbb{R}^2 \to \mathbb{R}^2$ which satisfy the following properties:

(1) $F(0) = 0$;
(2) $tr JF(P) < 0$ for all $P \in \mathbb{R}^2$;
(3) $det JF(P) > 0$ for all $P \in \mathbb{R}^2$.

For an example of a polynomial function in $\mathcal{F}$ which is not linear, see Exercise 4.5.3. We may now state the result of Meisters and Olech, [Meisters and Olech **88**, Theorem 1].

1.5 THEOREM. *Let F be a polynomial function in $\mathcal{F}$ then the origin is a globally asymptotically stable point of the system $\dot{X} = F(X)$.*

One of the key lemmas in the proof of this result has a purely D-module theoretic proof due to van den Essen, which we discuss in §2. Before we close this section, let us see how Theorem 1.5 can be applied to the Jacobian conjecture.

1.6 PROPOSITION. *Suppose that the origin is a global asymptotically stable point of $\dot{X} = F(X)$ for every polynomial map $F \in \mathcal{F}$. Then the polynomial maps in $\mathcal{F}$ are injective.*

PROOF: Suppose, by contradiction, that F is not injective. Then there exist points $P_1, P_2 \in \mathbb{R}^2$ such that $F(P_1) = F(P_2) = Q$. Consider the system $\dot{X} = H(X)$ where $H(X) = F(X + P_1) - Q$. Note that it has two distinct critical points, one at the origin and one at $P_2 - P_1 \neq 0$. Thus the origin cannot be globally asymptotically stable. However, $J(H)(X) = JF(X + P_1)$, and so H is a polynomial map in $\mathcal{F}$, which contradicts the hypothesis.

Since, by Theorem 1.5, the hypothesis in Proposition 1.6 is always satisfied by maps in $\mathcal{F}$, we conclude that these maps are always invertible. This is especially interesting since S. Pinchuk has recently given an example of a polynomial map on the plane whose determinant is everywhere positive, but which does *not* have an inverse, see [Pinchuk **94**].

2. GLOBAL UPPER BOUND

In this section we prove one of the key lemmas used to settle Theorem 1.5. The proof we give, due to van den Essen, is purely D-module theoretic and works over any field of characteristic zero. In this section we will make free use of the results of Ch. 4, §4.

Let $F : K^n \to K^n$ be a polynomial map and denote by $F_1, \ldots, F_n$ its coordinate functions. Let $\Delta(x) = det JF(x)$. Throughout this section we will assume that $\Delta(x) \neq 0$ for every $x \in K^n$. Note that since we are not assuming that K is algebraically closed, this does not imply that $\Delta(x)$ is constant. Put $d = \deg F = max\{\deg F_i : 1 \leq i \leq n\}$.

Let $g \in K[X, \Delta^{-1}]$ and consider the derivations D_i of $K[X, \Delta^{-1}]$ defined for $i = 1, \dots, n$ by

$$D_i(g) = \Delta^{-1} detJ(F_1, \dots, F_{i-1}, g, F_{i+1}, \dots, F_n)$$

as in Ch. 4, §4. Note that

(2.1) $$D_i = \Delta^{-1} \sum_1^n a_{ik} \partial_k$$

where a_{ik} is the ik cofactor. Hence, $\deg(a_{ik}) \leq (n-1)d$. By Lemma 4.4.1, these derivations satisfy

$$[D_i, F_j] = \delta_{ij} \text{ and } [D_i, D_j] = 0$$

for $1 \leq i, j \leq n$.

We shall use the D_i to define an A_n-module structure on $K[X, \Delta^{-1}]$ as follows:

$$x_i \bullet g = F_i \cdot q,$$
$$\partial_i \bullet g = D_i(q),$$

where $q \in K[X, \Delta^{-1}]$. A routine argument using Appendix 1 shows that $\bullet$ gives a well-defined action of A_n-module on $K[X, \Delta^{-1}]$. We denote this module by $M(F)$.

LEMMA 2.2. *As an A_n-module, $M(F)$ is holonomic and its multiplicity cannot exceed $2^n(2nd+1)^n$.*

PROOF: The proof follows the argument of Theorem 10.3.2. For $v \in \mathbb{N}$, put

$$\Gamma_v = \{g \cdot \Delta^{-2v} \in K[X, \Delta^{-1}] : \deg(g) \leq 2v(2nd+1)\}.$$

We show that $\{\Gamma_v\}_{v \geq 0}$ is a filtration of $M(F)$.

Let us show first that $B_i \cdot \Gamma_v \subseteq \Gamma_{v+i}$. It is enough to prove this for $i = 1$, because $B_i = B_1^i$. Let $q = g\Delta^{-2v} \in \Gamma_v$. Using the chain rule, we have that

$$\partial_i \bullet q = D_i(g)\Delta^{-2v} + (-2v)g\Delta^{-(2v+1)}D_i(\Delta).$$

Substituting for D_i the formula in (2.1) we get

$$\partial_i \bullet q = \Delta^{-2(v+1)} \left(\Delta \sum_1^n a_{ik} \partial_k(g) - 2vg \sum_1^n a_{ik} \partial_k(\Delta) \right).$$

Since $\deg(\Delta) \leq nd$ and $\deg(a_{ik}) \leq (n-1)d$ we conclude that $\partial \bullet q \in \Gamma_{v+1}$. A similar argument shows that $x_i \bullet q \in \Gamma_{v+1}$.

Finally we show that $\bigcup \Gamma_v = M(F)$. If $q \in K[X, \Delta^{-1}]$ then $q = g\Delta^{-r}$ where $g \in K[X]$ has degree s and $r \geq 0$. Put $v = max\{r, s\}$. Thus $q = g(\Delta^{2v-r})\Delta^{-2v}$ and since $v \geq s$,

$$\deg(g\Delta^{2v-r}) \leq s + (2v - r)nd \leq s + 2vnd \leq 2v(2nd + 1).$$

Thus $g\Delta^{-r} \in \Gamma_v$, which proves that $\{\Gamma_v\}_{v\geq 0}$ is a filtration of $M(F)$.

On the other hand, Γ_v is a K-vector space of dimension equal to that of the subspace of $K[X]$ of polynomials of degree $\leq 2v(2nd + 1)$. Hence,

$$\dim_K \Gamma_v \leq \frac{2^n(2nd + 1)^n}{n!} v^n + \text{terms of smaller degree in } v.$$

By Lemma 10.3.1, the module $M(F)$ is holonomic.

We are now ready to prove the main theorem of this section. Its purpose is to give a global bound on the number of elements in the inverse image $F^{-1}(P)$ of a point $P \in K^n$. Let $P = (P_1, \ldots, P_n)$. It is easy to see that the number of elements in $F^{-1}(P)$ equals the number of solutions of the system $F_1 - P_1 = \cdots = F_n - P_n = 0$.

2.3 THEOREM. *Let* $F : K^n \to K^n$ *be a polynomial map. If* $detJ(F) \neq 0$ *everywhere in* K^n, *then there exists a positive integer* b *such that* $F^{-1}(P)$ *does not have more than* b^n *points for every* $P \in K^n$.

PROOF: Let $P \in K^n$ and consider the polynomial map $F - P$. Since $J(F - P) = J(F)$, we have that $\Delta = detJ(F - P) = detJ(F) \neq 0$ everywhere in K^n. Put $M(P) = M(F - P)$. Note that $M(P) = K[X, \Delta^{-1}]$ for all $P \in K^n$, it is only the action of A_n on $M(P)$ that depends on P.

By Lemma 2.2, $M(P)$ is holonomic and its multiplicity is $2^n(2nd+1)^n = b$. By Theorem 18.1.4,

$$M(P)/ \sum_1^n (F_i - P_i)M(P) = M(P)/ \left(\sum_1^n x_i \bullet M(P) \right)$$

is a vector space over K of dimension $\leq b$. In particular, the classes of $1, x_1, x_1^2, \ldots, x_1^b$ in $M(P)/\sum_1^n x_i M(P)$ must be linearly dependent. Thus there exists a polynomial $g_1(x_1) \in K[x_1]$ of degree $\leq b$ and a positive integer r such that

$$(2.4) \qquad \Delta^r \cdot g(x_1) \in \sum_1^n (F_i - P_i) K[X].$$

Finally, if $Q = (Q_1, \ldots, Q_n) \in K^n$ satisfies $F(Q) = P$, then by (2.4) we have that $\Delta(Q)^r \cdot g(Q_1) = 0$. Since Δ has no zeros on K^n, it follows that $g(Q_1) = 0$. Hence there are at most b possibilities for the first coordinate of Q. Arguing similarly for the other coordinates we have that $F^{-1}(P)$ cannot have more than b^n elements. Since b is independent of P the theorem follows.

Theorem 2.3 and its proof are due to A. van den Essen [van den Essen **91**]. In the special case $K = \mathbb{R}$ this result follows from topological arguments; see [Bochnak, Coste and Roy **87**, Theorem 11.5.2].

3. GLOBAL STABILITY ON THE PLANE

We may now conclude the story that began in §1, by proving Theorem 1.5 on the global asymptotical stability of polynomial systems on the plane. The main ingredients of the proof are Theorem 2.3 and the following result.

THEOREM 3.1. *Let $F \in \mathcal{F}$. If there exist positive constants ρ and r such that*

$$|F(X)| \geq \rho \text{ whenever } |X| \geq r$$

then the origin is a globally asymptotically stable point of the system $\dot{X} = F(X)$.

Since the proof of this Theorem is purely analytic and not very straightforward, we will not give it here. It was first proved in [Olech **63**] where it follows from an application of Green's theorem. See also [Gasull, Libre and Sotomayor **91**]. Let us show how Theorems 2.3 and 3.1 can be used to prove Theorem 1.5.

PROOF OF THEOREM 1.5: Since $F \in \mathcal{F}$ it follows from Theorem 2.3 that

$$Sup\{\#F^{-1}(Y) : Y \in \mathbb{R}^2\} = K < \infty.$$

Let $P \in \mathbb{R}^2$ be a point at which the maximum is attained: that is $\#F^{-1}(P) = K$. Let $Q_1, \ldots, Q_K$ be the elements of $F^{-1}(P)$. By the inverse function theorem F is invertible in the neighbourhood of every point of $\mathbb{R}^2$. Hence for $1 \leq i \leq K$, it is possible to choose $\rho > 0$ and a neighbourhood V_i of Q_i such that

$$F : V_i \to B_\rho(P)$$

is a diffeomorphism, where $B_\rho(P)$ is the open ball centred on P of radius ρ. By decreasing the value of ρ, if necessary, we may also assume that $V_i \cap V_j = \emptyset$ if $i \neq j$. Let us prove that, under these hypotheses,

$$(3.2) \qquad F^{-1}(B_\rho(P)) = V_1 \cup \cdots \cup V_K.$$

It is clear that the union of the V_i's is contained in $F^{-1}(B_\rho(P))$. We prove the opposite inclusion. Suppose, by contradiction, that it does not hold. Thus there exists a point W not in $V_1 \cup \cdots \cup V_K$ such that $F(W) \in B_\rho(P)$. Since $F(V_i) = B_\rho(P)$, there are points $Y_i \in V_i$ such that

$$F(Y_i) = F(W)$$

for $i = 1, \ldots, K$. Note that if $i \neq j$ then $Y_i \neq Y_j$, because $V_i \cap V_j = \emptyset$. Furthermore $Y_i \neq W$ since W does not belong to the union $V_1 \cup \cdots \cup V_K$. Hence,

$$\{W, Y_1, \ldots, Y_K\} \subseteq F^{-1}(F(W)).$$

Thus $F^{-1}(F(W))$ cannot have less than $K + 1$ elements, a contradiction. Thus (3.2) holds.

Now choose $r' > 0$ so large that $B_{r'}(0)$ contains $V_1 \cup \cdots \cup V_K$. For this r' and the previously chosen ρ we have that

$$(3.3) \qquad |F(X) - P| \geq \rho \text{ if } |X| \geq r'.$$

Consider the translated function $G(X) = F(X + Q_1) - P$. It follows from (3.3) that $G(X)$ satisfies the hypothesis of Theorem 3.1 for $r = r' + |Q_1|$. Hence the system $\dot{X} = G(X)$ has the origin as a globally asymptotically stable point. Thus by Proposition 1.6 the map G is injective. But for $i = 1, \ldots, K$,

$$G(Q_i - Q_1) = F(Q_i) - P = 0 = G(0).$$

Since G is injective, we must have that $K = 1$. This means that $F^{-1}(P)$ has at most one point for any $P \in \mathbb{R}^2$. Thus (3.3) is satisfied by $P = 0$. But this is the hypothesis of Theorem 3.1, from which Theorem 1.5 immediately follows.

4. EXERCISES

4.1 Let A be a 2×2 matrix with real coefficients. Show that the origin is a globally asymptotically stable point of the system $\dot{X} = A \cdot X$ if and only if the real part of the eigenvalues of A are negative.

4.2 Let $F \in \mathcal{F}$. Show that if F is globally invertible in $\mathbb{R}^2$ then the origin is a globally asymptotically stable point of the system $\dot{X} = F(X)$.

Hint: By the inverse function theorem there exist $\rho, r > 0$ such that F maps $B_\rho(0)$ into $B_r(0)$. Since F is globally one-to-one the points outside $B_\rho(0)$ must be sent outside $B_r(0)$. But this is the hypothesis of Theorem 3.1.

4.3 Let $F \in \mathbb{R}[x, y]$. Use Green's theorem to show that if $detJ(F) = 1$ everywhere on $\mathbb{R}^2$ then F is a map of $\mathbb{R}^2$ that preserves area.

CHAPTER 20

AUTOMATIC PROOF OF IDENTITIES

Special functions are back in fashion. Recent years have seen the development of new approaches to the theory and also of many applications of special function identities. Two important examples of the latter are de Branges' proof of Bieberbach's conjecture, [de Branges **85**], and Apéry's proof of the irrationality of $\zeta(3)$, [van der Poorten **79**]. Holonomic modules can be used to find the differential equations satisfied by certain special functions and also to determine whether a given identity is true. In many cases this can be done automatically; that is, by a computer. In this chapter we give an introduction to the theoretical foundations of this approach, pioneered by Zeilberger and his collaborators.

1. HOLONOMIC FUNCTIONS.

We will assume throughout this chapter that the coefficient field is $\mathbb{R}$. Thus A_n will stand for $A_n(\mathbb{R})$. A function is holonomic if it is the solution of a holonomic module. Let U be an open set of $\mathbb{R}^n$. If $f \in C^\infty(U)$, set

$$I_f = \{d \in A_n(\mathbb{R}) : d(f) = 0\}.$$

Then f is *holonomic* if A_n/I_f is a holonomic module. But A_n/I_f is isomorphic to the submodule $A_n f$ of $C^\infty(U)$ generated by f. Thus f is a holonomic function if and only if $A_n f$ is a holonomic module.

A polynomial $f \in \mathbb{R}[x_1, \ldots, x_n]$ is a holonomic function. Indeed, if f has total degree k, then I_f contains the monomials ∂^α with $|\alpha| = k + 1$. Thus A_n/I_f is holonomic, see Exercise 4.1. The next result is of great help in producing examples of holonomic functions.

1.1 PROPOSITION. *Let $f, g \in C^\infty(U)$ be holonomic functions. Then $f + g$ and fg are holonomic.*

PROOF: Since f and g are holonomic functions, the modules $A_n f$ and $A_n g$ are holonomic. Hence $A_n f \oplus A_n g$ and its subquotient $A_n(f+g)$ are holonomic, by Proposition 10.1.1. Thus $f + g$ is a holonomic function.

Let $\mathcal{M}$ be the A_n-submodule of $C^\infty(U)$ generated by $\partial^\alpha(f)\partial^\beta(g)$ for all $\alpha,\beta \in \mathbb{N}^n$. It follows from Leibniz's rule that $A_n(fg) \subseteq \mathcal{M}$. Now, the multiplication map defines a homomorphism of A_n-modules,

$$A_nf \otimes_{\mathbb{R}[X]} A_ng \to C^\infty(U),$$

whose image is $\mathcal{M}$. Since $A_nf \otimes_{\mathbb{R}[X]} A_ng$ is holonomic by Exercise 18.4.1, it follows that $A_n(fg)$ is holonomic by Proposition 10.1.1.

The composition of holonomic functions, however, is *not* holonomic. As we have seen in Ch. 5, §3, the function $\exp(\exp(x))$ does not satisfy a differential equation with polynomial coefficients. Hence, although the exponential function is holonomic, the composite $\exp(\exp(x))$ is not holonomic.

2. HYPEREXPONENTIAL FUNCTIONS.

Let $a_1, a_2, \ldots$ be a sequence of real numbers. It is called a *geometric* sequence, if the sequence of ratios a_{n+1}/a_n is constant. A natural generalization is to assume that the quotients a_{n+1}/a_n are a rational function of n. These sequences are called *hypergeometric*.

In the realm of functions, the object that corresponds to a geometric sequence is the exponential function, which satisfies: f'/f is constant. This suggests that a function $f \in C^\infty(U)$ should be called *hyperexponential* if $\partial_i(f)/f$ is a rational function of $x_1, \ldots, x_n$ for $i = 1, \ldots, n$. Note that if q is a rational function in $\mathbb{R}(X)$, then $\exp(q)$ is hyperexponential. To produce more examples, one may use the next result.

2.1 PROPOSITION. *The product of hyperexponential functions is hyperexponential.*

PROOF: Let f, g be hyperexponential functions. By Leibniz's rule,

$$\frac{\partial_i(fg)}{fg} = \frac{\partial_i(f)}{f} + \frac{\partial_i(g)}{g}.$$

The result follows from the fact that the sum of rational functions is a rational function.

We will now show that hyperexponential functions are holonomic. This is easy in one variable. If f is hyperexponential, then $\partial(f)/f = p/q$, where p, q are polynomials. Thus f satisfies the differential equation

$$(q\partial - p)(f) = 0.$$

Equivalently, f is a solution of the module $A_1/A_1(q\partial - p)$. Since $q \neq 0$ and we are in dimension 1, this module is holonomic. Thus f is holonomic.

The n-dimensional case is very much harder. Suppose that f is a hyperexponential function of n variables. Then $\partial_i(f)/f = p_i/q_i$, where $p_i, q_i \in \mathbb{R}[X]$. Thus f satisfies the system of differential equations

$$(q_i\partial_i - p_i)(f) = 0$$

for $1 \leq i \leq n$. Let $J = \sum_1^n A_n(q_i\partial_i - p_i)$. Then $J \subseteq I_f$.

Put $M = A_n/J$ and let q be the least common multiple of $q_1, \ldots, q_n$. Denote by $M[q^{-1}]$ the module $\mathbb{R}[X, q^{-1}] \otimes_{\mathbb{R}[x]} M$; see Ch. 12, §5. Every element of $M[q^{-1}]$ can be written in the form $q^{-k} \otimes u$, where $u \in M$. Recall that the action of ∂_i on $q^{-k} \otimes u \in M[q^{-1}]$ is defined by

$$\partial_i(q^{-k} \otimes u) = -kq^{-k-1}\partial_i(q) \otimes u + q^{-k} \otimes \partial_i u.$$

2.2 THEOREM. *If $M[q^{-1}]$ is holonomic, then f is holonomic.*

PROOF: Consider the map

$$\phi: A_n/I_f \to \mathbb{R}[X, q^{-1}] \otimes A_n/I_f$$

defined by $\phi(d+I_f) = 1 \otimes (d+I_f)$. By Proposition 12.5.1, the element $d+I_f$ is in the kernel of ϕ if and only if $q^k d \in I_f$ for some positive integer k. This happens if $q^k d(f) = 0$. Since q is a polynomial, we must have that $d(f) = 0$. Hence $d \in I_f$. Summing up: the homomorphism ϕ is injective.

Now, since $J \subseteq I_f$, it follows that A_n/I_f is a homomorphic image of M. Thus $\mathbb{R}[X, q^{-1}] \otimes A_n/I_f$ is a homomorphic image of $M[q^{-1}]$, by Theorem 12.4.6. Hence $\mathbb{R}[X, q^{-1}] \otimes A_n/I_f$ is holonomic over A_n. But we have seen that A_n/I_f is a submodule of $\mathbb{R}[X, q^{-1}] \otimes A_n/I_f$, thus it is itself holonomic. Hence f is a holonomic function.

Let us prove that $M[q^{-1}]$ is a holonomic A_n-module. We begin with a technical lemma.

2.3 LEMMA. *Let* $s = \max\{\deg(p_i) : 1 \leq i \leq n\} + \deg(q)$. *If* d *is an operator of* A_n *of degree* k, *then*

$$q^k d \equiv g \pmod{J}$$

where g *is a polynomial of degree* $\leq ks$.

PROOF: The operator d is a *finite* linear combination of monomials $x^\alpha \partial^\beta$ and q commutes with x^α. Thus we have only to prove the following statement: if $\beta \in \mathbb{N}^n$ and $|\beta| = k$, then $q^k \partial^\beta \equiv g \pmod{J}$ where $g \in \mathbb{R}[X]$ has degree $\leq ks$.

We will proceed by induction on k. If $k = 1$, then $\partial^\beta = \partial_i$ for some $i = 1, \ldots, n$. Note that since q is the least common multiple of $q_1, \ldots, q_n$, it follows that q/q_i is a polynomial. Thus,

$$q\partial_i \equiv p_i(q/q_i) \pmod{J}.$$

and $\deg(p_i(q/q_i)) \leq \deg(p_i) + \deg(q/q_i) \leq s$.

Assume that the result holds for $k-1$ and let us prove that $q^k \partial^\beta \in \mathbb{R}[X]+J$ when $|\beta| = k \geq 2$. Since $|\beta| > 0$, it follows that $\beta_i \neq 0$ for some $i = 1, \ldots, n$. Now

(2.4) $$q^k \partial^\beta = q^k \partial_i(\partial^{\beta - e_i}).$$

But

$$q^k \partial_i = \partial_i \cdot q^k - kq^{k-1} \partial_i(q).$$

By the induction hypothesis there exists $g \in \mathbb{R}[X]$ of degree $\leq (k-1)s$ such that $q^{k-1} \partial^{\beta - e_i} \equiv g \pmod{J}$. Substituting in (2.4) we obtain

$$q^k \partial^\beta \equiv \partial_i \cdot q \cdot g - kg\partial_i(q) \pmod{J}.$$

But $\partial_i \cdot qg = g(q\partial_i) + \partial_i(qg)$. Since $q\partial_i \equiv p_i(q/q_i) \pmod{J}$, we conclude that

$$q^k \partial^\beta \equiv gp_i(q/q_i) + \partial_i(gq) - kg\partial_i(q) \pmod{J}.$$

The right hand side is a polynomial of degree less than or equal to

$$\max\{(k-1)s + s, (s(k-1) + s) - 1, s(k-1) + (s-1)\} \leq sk$$

as required.

The proof of the theorem follows the argument already used in Theorem 10.3.2 and Ch. 12, §5.

2.5 THEOREM. *The A_n-module $M[q^{-1}]$ is holonomic.*

PROOF: Let $m = \deg(q)$ and Γ be the good filtration of M induced by the Bernstein filtration of A_n. Define

$$\Omega_k = \{q^{-k} \otimes u : u \in \Gamma_{(m+1)k}\}.$$

This is a filtration of $M[q^{-1}]$ as shown in Theorem 12.5.4. By Lemma 2.3,

$$q^{-k} \otimes u = q^{-k(m+2)} \otimes q^{k(m+1)}u = q^{-k(m+2)} \otimes g,$$

where g is a polynomial of degree $\leq s(m+1)k$. Thus, as a real vector space,

$$dim\Omega_k \leq \binom{s(m+1)k + n}{n}.$$

By Lemma 10.3.1, it follows that $M[q^{-1}]$ is holonomic.

Putting together Theorems 2.2 and 2.5 we have the required result.

2.6 THEOREM. *A hyperexponential function in n variables is holonomic.*

A more general result can be found in [Takayama **92**, Appendix]. In the next section we explain the theoretical foundations of an algorithm which calculates the differential equation satisfied by a given definite integral with parameters. As an example, we calculate the integral of a hyperexponential function.

3. THE METHOD.

Denote by x, y the coordinate functions of $\mathbb{R}^2$. Let $U = (a, b) \times (-\infty, \infty)$ and f be a function in $C^\infty(U)$ which satisfies

(3.1) $$\lim_{y \to \pm\infty} x^\alpha \partial^\beta f(x, y) = 0$$

for all $\alpha, \beta \in \mathbb{N}^2$. This implies that if $P \in A_2$, then

$$\int_{-\infty}^{\infty} P(f)dy < \infty.$$

for every $x \in (a, b)$. In particular this is the case for $P = 1$. Put

$$R(x) = \int_{-\infty}^{+\infty} f(x, y) \, dy.$$

We will describe an algorithm which finds a differential equation satisfied by $R(x)$, when f is a given holonomic function.

Let $M = A_2 / I_f$ and $\pi : \mathbb{R}^2 \to \mathbb{R}$ be the projection on the first coordinate. The direct image of M under π is

$$\pi_* M \cong M / \partial_y M \cong A_2 / (I_f + \partial_y A_2).$$

If we now assume that f is a holonomic function, then M is a holonomic A_2-module. Thus $\pi_* M$ is holonomic over A_1 by Theorem 18.2.2. Hence the kernel of the homomorphism

$$A_1 \to A_2 / (I_f + \partial_y A_2)$$

which maps $d \in A_1$ to $d + (I_f + \partial_y A_2)$ must be non-zero. Thus there exists a non-zero operator $D \in A_1$ such that

$$D = Q_1 + \partial_y \cdot Q_2$$

where $Q_1 \in I_f$ and $Q_2 \in A_2$. Since $Q_1(f) = 0$, we have that

(3.2) $D(f) = \partial_y Q_2(f).$

Integrating the right hand side of (3.2) between $-\infty$ and ∞ and using the fundamental theorem of calculus,

$$\int_{-\infty}^{\infty} \partial_y Q_2(f) \, dy = Q_2(f) \big|_{-\infty}^{+\infty} = 0.$$

Thus, integrating the left hand side of (3.2) and using differentiation under the integral sign [Buck **56**, Ch. 4, §4.4, Theorem 29], one gets

$$0 = \int_{-\infty}^{\infty} D(f) \, dy = D(R).$$

Summing up: $R(x)$ satisfies the differential equation $D(R) = 0$.

We will apply the method to an example, borrowed from [Almkvist and Zeilberger 90]. Let $f(x) = \exp(-(x/y)^2 - y^2)$ and $R(x)$ be its integral between $-\infty$ and ∞. The function f is hyperexponential, hence holonomic by Theorem 2.6. It is easy to check that it satisfies condition (3.1). A simple calculation shows that f is a solution of the equations

$$(y^2 \partial_x + 2x)(f) = 0,$$
$$(y^3 \partial_y + 2y^4 - 2x^2)(f) = 0.$$

Let $L = I_f + \partial_y A_2$. Since $y^3 \partial_y = \partial_y \cdot y^3 - 3y^2$, we have that

$$-3y^2 + 2y^4 \equiv 2x^2 \quad (\text{mod } L).$$

We are allowed to multiply this identity by ∂_x^3 on the left, because L is an A_1-submodule of A_2. Doing this and using the identity $y^2 \partial_x \equiv -2x \quad (\text{mod } J)$, three times, we get that

$$D = 6\partial_x \cdot x + 8\partial_x \cdot x^2 + 8x - 2\partial_x^3 \cdot x^2$$

is in L. Now D factors as

$$(x\partial_x + 3)(\partial_x - 2)(\partial_x + 2).$$

In this special case we may use the differential equation $D(R) = 0$ to calculate R. The general solution of this equation is of the form

$$c_1 \exp(2x) + c_2 \exp(-2x) + c_3 h,$$

where h satisfies the differential equation

$$d^2 h/dx^2 - 4h = Ax^{-3}$$

for some constant $A \neq 0$.

We must now decide, among the possible solutions, which one coincides with $R(x)$. First of all, h cannot be bounded at 0. But

$$R(0) = \int_{-\infty}^{\infty} \exp(-y^2) dy = (\sqrt{\pi})^{-1};$$

thus $c_3 = 0$. Furthermore, $\lim_{x \to +\infty} R(x) = 0$, hence $c_1 = 0$. We are left with $R(x) = c_2 \exp(-2x)$, and so

$$(\sqrt{\pi})^{-1} = R(0) = c_2.$$

Therefore, $R(x) = (\sqrt{\pi})^{-1} \exp(-2x)$. This is to be taken as a mere illustration: no one would dream of calculating this particular integral in this way. For the easy solutions, see Exercise 4.5.

In the above example we found D from the equations for f by an *ad hoc* calculation. Takayama showed in [Takayama **92**] that one may use Gröbner bases to determine D. If f is hyperexponential, there is also an algorithm in [Almkvist and Zeilberger **90**], based on Gosper's summation algorithm [Gosper **78**]. Thus the calculation of D can be done automatically.

If one tries to extend the algorithm to find $R(x)$ by solving $D(R) = 0$ like we did, the difficulties are of another order of magnitude. There are two main problems:

(1) The integration $D(R) = 0$.

(2) Checking initial conditions.

One way to get around (1) is to settle for an algorithm to *certify* an identity. In other words, we have 'guessed' $R(x)$ by whatever method, and we want to show that our guess is correct. That still leaves (2). Since the equation $D(R) = 0$ is usually of order greater than 1, checking initial conditions can be very tricky.

However, hyperexponential functions are like hypergeometric sequences, and integrals like sums; and it turns out that these ideas can be used to certify identities of hypergeometric sums. In this case (2) does not present any problem and the algorithm is very effective. However, we are no longer dealing with modules over the Weyl algebra. The base ring in this case is a *localization* of the Weyl algebra. Many interesting applications of these algorithms are discussed in [Zeilberger **90**], [Almkvist and Zeilberger **90**] and [Cartier **92**].

4. EXERCISES

4.1 Let J_k be the left ideal of A_n generated by the monomials ∂^α with $|\alpha| = k$. Show that A_n/J_k is a holonomic A_n-module.

Hint: There exists an exact sequence,

$$0 \to J_k/J_{k+1} \to A_n/J_{k+1} \to A_n/J_k \to 0$$

and $J_k/J_{k+1} \cong K[X]$.

4.2 Show that the derivative of a hyperexponential function is hyperexponential.

4.3 Find an upper bound for the multiplicity of the module A_n/I_f, when f is a hyperexponential function.

Hint: What is the bound on the multiplicity of $M[q^{-1}]$ in Theorem 2.5?

4.4 Show that the function $\sin(xy)$ is holonomic, but *not* hyperexponential.

4.5 Let $f(x, y) = \exp(-(x/y)^2 - y^2)$ and $R(x)$ be the integral of f with respect to y between $-\infty$ and ∞. Calculate $R(x)$ in two different ways, as follows:

   (1) Using the substitution $t = y - x/y$ in $\int_{-\infty}^{\infty} \exp(-t^2)dt = (\sqrt{\pi})^{-1}$.
   (2) Using differentiation under the integral sign to obtain the equation $R' = -2R$ and integrating it.

4.6 Let $f(x, y) = \exp(-xy^2)$ and $R(x) = \int_{-\infty}^{\infty} f(x, y)dy$. Find an operator $D \in A_1$ such that $D(R) = 0$.

4.7 Let $\lambda_1, \ldots, \lambda_s \in \mathbb{R}^n$ and let X be the n-tuple $(x_1, \ldots, x_n)$. Denote by $\lambda_i \cdot X$ the formal inner product of the two n-tuples. For $p_1, \ldots, p_s \in K[x_1, \ldots, x_n]$ put

$$f(x) = \sum_{1}^{s} p_i(X)\exp(\lambda_i \cdot X).$$

Show that f is a holonomic function.

CODA

Since this book is only a primer, it is convenient to give the interested reader directions for further study. The comments that follow are based on this author's experience and inevitably reflect his tastes.

First of all, the theory of algebraic D-modules is itself a part of algebraic geometry. Thus we must start with an algebraic variety X. If we assume that X is affine, then its algebraic geometric properties are coded by the ring of polynomial functions on X (and its modules). This is a commutative ring, called the *ring of coordinates* and denoted by $O(X)$. The ring of differential operators $D(X)$ is the ring of differential operators of $O(X)$ as defined in Ch. 3. If the variety is smooth (non-singular) then $D(X)$ is a simple noetherian ring.

To deal with general varieties it is necessary to introduce sheaves. The *structure sheaf* keeps the same relation to a general variety as the coordinate ring does to an affine variety. From it we may derive the *sheaf of rings of differential operators*. If the variety is smooth, this is a coherent sheaf of rings. The purpose of D-module theory is the study of the category of coherent sheaves of modules over the sheaf of rings of differential operators of an algebraic variety.

It is plain that a good knowledge of algebraic geometry is essential to make sense of these statements. The standard reference is the first three chapters of [Hartshorne **77**]. One can also find the required sheaf theory in Serre's beautiful "Faisceaux algébriques cohérents", [Serre **55**]. But a thorough grounding in classical algebraic geometry is necessary before one tackles this paper. Two good references for that are [Harris **92**] and the first chapter of [Hartshorne **77**].

With this background it is possible to extend most of the material covered in the book to general varieties, including a full version of Kashiwara's theorem. Unfortunately there is no introduction to algebraic D-modules assuming only a good knowledge of sheaf theory. In this respect its twin sister, the theory of analytic D-modules, is better served. Analytic D-modules have an *analytic* (instead of algebraic) manifold for base space. The two theories are very close, but there are some differences. However, with a little care one should be able to use a book on analytic D-modules to learn about the algebraic theory. An excellent introduction is [Pham **79**]. A more recent work in the same spirit is [Granger and Maisonobe **93**]. See also [Björk **79**]. Since we have mentioned family relationships: D-module theory has another close relative in the theory of modules over rings of microdifferential operators; see [Schapira **85**].

It is always good to have a substantial result to aim at when one is first learning a theory. There is Kashiwara's theorem, of course. Another result that one could tackle is the theorem of Beilinson and Bernstein on the structure of modules over the ring of differential operators of projective space: [Beilinson and Bernstein **81**]. This has the added advantage that the theorem was one of the main steps in the proof of a famous conjecture in representation theory, see [Kirwan **88**]. The Beilinson-Bernstein theorem is proved in [Borel et al. **87**, Ch. VII §8] and [Benoist **93**].

The background so far described is not enough when one comes to study direct images and the structure theory of holonomic modules. For example, direct images can be easily generalized to varieties following the principles of Ch. 16, but this definition will not produce a functor. The problem is that Theorem 18.2.1 will break down for general varieties. To get over this problem it is necessary to give up the categories of sheaves of D-modules and work instead with their derived categories.

Derived categories belong to homological algebra. It turns out that homological algebra is, right from the beginning, one of the essential tools in the study of D-modules. In fact, to avoid using homological algebra we had to make great sacrifices in this book. One obvious point has to do with the duality properties of holonomic modules, which are defined using the functor *Ext*, see [Borel et al. **87**, Ch. V] and [Milicic **86**].

Unfortunately it is not possible to come to a deep understanding of D-modules without derived categories. This is a very technical subject. A good starting point is [Iversen **86**]. For a more detailed approach see [Borel et al. **87**, Ch. I]. The ultimate introduction to algebraic D-modules with the full benefit of derived categories is [Borel et al. **87**]: it goes as far as the

Riemann-Hilbert correspondence, one of the jewels of this theory.

A lot of background is required before one gets to the heart of D-module theory, which means a rather long preparation. Is it worth it? The answer must be yes: the ascent may be steep, but the view is truly breathtaking. Besides, the power of this machinery is such that applications are legion and contacts with other areas of mathematics the rule.

APPENDIX 1
DEFINING THE ACTION OF A MODULE
USING GENERATORS

In Ch. 1 we defined the Weyl algebra in two different ways. First as a subalgebra generated by certain endomorphisms of the polynomial ring. Second as the quotient of a free algebra by an ideal of relations. In both cases, an arbitrary element of the Weyl algebra is only determined *a posteriori*, as a linear combination of monomials. on the generators. So when it comes to defining an action of a Weyl algebra on a vector space, it should be enough to say how the generators are to act on the vectors. This would be the case if the generators were not bound up by relations. It is the relations that dictate which actions are well-defined and which are not. This is a very simple matter, but it is usually not carefully discussed in elementary books. It plays such an important rôle in this book that it seems better to treat it in detail in this appendix.

We begin with a very general setup. Let R be a K-algebra and let M be a left R-module. The action of R on M gives rise to a homomorphism of K-algebras

$$\phi : R \longrightarrow End_K(M)$$

which maps an element $a \in R$ onto the endomorphism ϕ_a of M defined by $\phi_a(u) = au$, for $u \in M$. Given such a homomorphism we may easily recover the action of R by the preceding formula.

Suppose now that F is a free algebra and that there is a surjective homomorphism $\pi : F \to R$. The composition $\phi \cdot \pi$ gives a map from F to $End_K(M)$ that makes M into an F-module. Besides $Ker(\pi)M = 0$. This last equation tells us that M is not only an F- module, but in fact an R-module.

The recipe for checking whether a proposed action of generators is well-defined is inspired by the setup above. Assume now that the algebra R is generated by $a_1, \ldots, a_k$. Let F be the free algebra on k generators $x_1, \ldots, x_k$ and let M be a K-vector space. Suppose that an action $a_i u$ has been postulated for every $u \in M$ and $i = 1, \ldots, k$. How can we check whether this is truly an action of R on M?

First of all put $x_i u = a_i u$. In this way we have defined an action of F (and not just its generators) on M, because the generators of F do *not* satisfy any relations. Hence any arbitrary definition will give a good action. Thus we have a map $\psi : F \longrightarrow End_K(M)$ such that

$$\psi_{x_i}(u) = x_i u = a_i u,$$

for every $u \in M$. There is also a map $\pi : F \longrightarrow R$, given by $\pi(x_i) = a_i$. We need to know whether ψ factors through π. This will happen if $Ker\pi \subseteq Ker\psi$. In other words, a well defined action of R on M has been defined if $Ker\pi \cdot M = 0$.

Let us apply this recipe to the Weyl algebra. Let M be a K-vector space. To define an action of $A_n(K)$ on M we begin by prescribing values for $x_i u$ and $\partial_i u$, for every $u \in M$ and $i = 1, \ldots, n$. Since the ideal of relations is generated by a finite number of elements, the next step is very easy. All we have to do is to make sure that

$$[\partial_i, x_j] u = \delta_{ij} u$$

and

$$[x_i, x_j] u = [\partial_i, \partial_j] u = 0$$

for every $u \in M$. This is not usually easy to do, as the following example shows. It is borrowed from Ch. 19, §2.

Let $F_1, \ldots, F_n$ be polynomials in $K[x_1, \ldots, x_n]$ and suppose that there exist pairwise commuting derivations $D_1, \ldots, D_n$ such that $D_i(F_j) = \delta_{ij}$. This will be the case if the jacobian determinant of $F_1, \ldots, F_n$ is 1; cf. Ch. 4, §4. We would like to define an action of A_n on $K[X]$ by the formulae:

$$x_i \bullet g = F_i g$$
$$\partial_i \bullet g = D_i(g)$$

where $g \in K[X]$. Is this action well defined? Let us check that one of the relations, say $[\partial_i, x_j] = \delta_{ij}$, is satisfied on $g \in K[X]$. Using the definitions, we have that

$$(\partial_i \cdot x_j) \bullet g = \partial_i \bullet (F_j g) = D_i(F_j g).$$

Since D_i is a derivation, we may apply Leibniz's rule to get

$$[\partial_i, x_j] \bullet g = D_i(F_j)g = \delta_{ij}g.$$

The other relations can be similarly checked, and we conclude that the action can be extended to all elements of A_n. The most important application of this principle occurs in the definition of inverse image in Ch. 14.

APPENDIX 2
LOCAL INVERSION THEOREM

The purpose of this appendix is to give a self-contained proof of the local inversion theorem, which is used in the proof of Lemma 4.4.1. Our exposition follows [Bourbaki 59].

LOCAL INVERSION THEOREM. *Let $F = (F_1, \ldots, F_n)$ be an n-tuple of power series in $K[[x_1, \ldots, x_n]]$. Assume that the jacobian determinant $\det JF(0) \neq 0$ and that $F(0) = 0$. Then there exist $G_1, \ldots, G_n \in K[[x_1, \ldots, x_n]]$, without constant terms, such that*

$$x_i = G_i(F_1, \ldots, F_n)$$

for $i = 1, \ldots, n$.

The proof of the theorem follows from the following lemma.

LEMMA. *For $1 \leq i \leq m$, let*

$$P_i(y_1, \ldots, y_m, x_1, \ldots, x_n) \in K[[y_1, \ldots, y_m, x_1, \ldots, x_n]]$$

be formal power series and denote by $P(Y, X)$ the corresponding m-tuple. Let

$$\Delta(Y, X) = \det\left(\frac{\partial P_i}{\partial y_j}\right)_{1 \leq i, j \leq m}.$$

Suppose that $P(0, 0) = 0$ and that $\Delta(0, X)$ is invertible in $K[[x_1, \ldots, x_n]]$. Then there exists a unique m-tuple $G = (G_1, \ldots, G_m)$ in $K[[x_1, \ldots, x_n]]$ such that

$$P_i(G_1, \ldots, G_m, x_1, \ldots, x_n) = 0$$

for $i = 1, \ldots, m$.

PROOF: We may write each P_i as a power series in the y's with coefficients in $K[[X]] = K[[x_1, \ldots, x_n]]$. Using multi-indices, we have that

$$P_i = a_{i0} + \sum_1^m a_{ij} y_j + \sum_{|\alpha| \geq 2} a_{i\alpha} y^\alpha,$$

where a_{i0}, a_{ij} and $a_{i\alpha}$ belong to $K[[X]]$ and a_{i0} has no constant term. We begin by showing that it is enough to prove the lemma in a special case.

By hypothesis, the matrix $(a_{ij})_{1 \leq i,j \leq m}$ is invertible over $K[[X]]$. Let us denote its inverse by $(b_{ij})_{1 \leq i,j \leq m}$. Put

$$Q_i = \sum_1^m b_{ij} P_j.$$

Then we have

$$Q_i = -c_{i0} + y_i - \sum_{|\alpha| \geq 2} c_{i\alpha} y^\alpha$$

where $c_{i0}, c_{i\alpha} \in K[[X]]$ and c_{i0} has no constant term. It is enough to solve the problem for the Q's, because $P_i = \sum_1^m a_{ij} Q_j$.

Suppose now that series $G_1, \ldots, G_m$ have been obtained which satisfy

$$Q_i(G_1, \ldots, G_m, x_1, \ldots, x_n) = 0.$$

Thus,

(1) $$G_i = c_{i0} + \sum_{|\alpha| \geq 2} c_{i\alpha} G^\alpha.$$

Let S_{ik} be the sum of the homogeneous components of degree $\leq k$ of G_{ik}. Given a multi-index $\alpha = (\alpha_1, \ldots, \alpha_m)$ we will write

$$S_k^\alpha = S_{1k}^{\alpha_1} \ldots S_{mk}^{\alpha_m}.$$

Since the G's have no constant term, the component of degree k of G^α, for $|\alpha| \geq 2$, coincides with the component of degree k of S_{k-1}^α. It follows from (1) that S_1 is equal to the homogeneous component of degree 1 of c_{i0}. Furthermore, for all $k > 1$ the homogeneous component of degree k of G_i coincides with the corresponding component of

$$c_{i0} + \sum_{|\alpha| \geq 2} c_{i\alpha} S_{k-1}^\alpha.$$

and can thus be determined by recursion. Note that since the coefficients of G_i are recursively determined using (1), it follows that G_i itself is unique.

Let us prove that the local inversion theorem follows from this lemma.

PROOF OF THE LOCAL INVERSION THEOREM: Denote by P the n-tuple of power series in $K[[y_1, \ldots, y_n, x_1, \ldots, x_n]]$ whose components are

$$P_i = F_i(y_1, \ldots, y_n) - x_i$$

for $1 \le i \le n$. Then $P(0,0) = 0$ and

$$\Delta(Y, X) = detJF(Y),$$

so that $\Delta(0, X) = detJF(0)$ is a non-zero constant. Hence we may apply the lemma, which asserts that there exist power series $G_1, \ldots, G_n \in K[[X]]$ such that

$$0 = P_i(G_1, \ldots, G_n, x_1, \ldots, x_n) = F_i(G_1, \ldots, G_n) - x_i$$

as claimed in the theorem.

REFERENCES

Almkvist, G. and Zeilberger, D.
1990 The method of differentiation under the integral sign, *J. Symbolic Computation*, **10**, 571-591.

Arnold, V.I.
1981 *Ordinary differential equations*, The MIT Press, Cambridge, Mass.-London.

Barabanov, N. E.
1988 On a problem of Kalman, *Siberian Math. J.*,**29**, 333-341.

Bass, H., Connell, E.H. and Wright, D.
1982 The Jacobian Conjecture: reduction of degree and formal expansion of the inverse, *Bull. Amer. Math. Soc.(New Series)*, **7**, 287-330.

Beilinson, A. and Bernstein, J.
1981 Localisation de g-modules, *C.R. Acad. Sci. Paris*, **292**, 15-18.

Benoist, Y.
1993 *D*-module sur la variété des drapeaux, in *Images directes et constructibilité*, ed. P. Maisonobe et C. Sabbah, Travaux en Cours, 46, Hermann, Paris.

Bernstein, I.N.
1971 Modules over a ring of differential operators: study of the fundamental solutions of equations with constant coefficients, *Funct. Anal. Appl.*, **5**, 89-101.

1972 The analytic continuation of generalized functions with respect to a parameter, *Funct. Anal. Appl.*, **6**, 273-285.

Bernstein J. and Lunts, V.
1988 On non-holonomic irreducible D-modules, *Invent. Math.*, **94**, 223-243.

Björk, J.-E.
1979 *Rings of differential operators*, North Holland Mathematics Library 21, Amsterdam.

Bochnak, J., Coste, M. and Roy, M.-F.
1987 *Géométrie algébrique réelle*, Ergebnisse der Math. U.i. Grenzgebieten, Springer-Verlag, Berlin-Heidelberg-New York.

Borel, A. et al.
1987 *Algebraic D-modules*, Perspectives in Mathematics 2, Academic Press, London-New York.

Bourbaki, N.
1959 *Algèbre, chapitre 4: polynômes et fractions rationelles*, deuxième édition, Actualités Scientifiques et Industrielles 1102, Hermann, Paris.

Briançon, J. and P. Maisonobe
1984 Idéaux de germes d'opérateurs différentiels à une variable, *Ens. Math.*, **30**, 7-38.

Buck, R.C.
1956 *Advanced calculus*, McGraw-Hill Book Company, New York-Toronto-London.

Cartier, P.
1992 Démonstrations "automatique" d'identités et fonctions hypergéométriques [d'après D. Zeilberger], Séminaire Bourbaki 1991/92, *Astérisque*, **206**, 41-91.

Cassou-Noguès, P.
1986 Racines de polynômes de Bernstein, *Ann. Inst. Fourier, Grenoble*, **36**, 1-30.

Cohn, P.M.
1979 *Algebra*, vol. 2, John Wiley and Sons, Chichester-New York-Brisbane-Toronto-Singapore.

1984 *Algebra*, vol. 1, second edition, John Wiley and Sons, Chichester-New York-Brisbane-Toronto-Singapore.

Coutinho, S.C. and Holland, M.P.
1988 Module structure of rings of differential operators, *Proc. London Math. Soc.*, **57**, 417-432.

de Branges, L.
1985 A proof of the Bieberbach conjecture, *Acta Math.*, **154**, 137-152.

Dirac, P.A.M.
1978 The development of quantum mechanics, in *Dirac: Directions in Physics*, ed. H. Hora and J.R. Shepanski, John Wiley and Sons, New York-London-Sidney-Toronto.

Dixmier, J.
1963 Représentations irréductibles des algèbres de Lie nilpotentes, *Anais Acad. Bras. Ciênc.*, **35**, 491-519.

1968 Sur les algèbres de Weyl, *Bull. Soc. Math. France*, **96**, 209-242.

1974 *Algèbres enveloppantes*, Cahiers Scientifiques 37, Gauthier-Villars, Paris.

Gabber, O.
1981 The integrability of the characteristic variety, *Amer. J. Math.*, **103**, 445-468.

Gasull, A., Llibre, J. and Sotomayor, J.
1991 Global asymptotic stability of differential equations in the plane, *J. Diff. Eq.*, **91**, 327-335.

Gosper Jr, R.W.
1978 Decision procedure of indefinite summation, *Proc. Nat. Acad. Sci. USA*, **75**, 40-42.

Granger, M. and Maisonobe, P.
1993 A basic course on differential modules, in *D-modules cohérents et holonomes*, ed. P. Maisonobe et C. Sabbah, Travaux en Cours 45, Hermann, Paris.

Gutierrez, C.
1993 A solution to the bidimensional global asymptotic stability conjecture, *Informes de Matemática IMPA*, Série A-097.

Hardy, G.H.
1928 *The integration of functions of a single variable*, Cambridge Tracts in Mathematics and Mathematical Physics 2, Cambridge University Press, London.

Harris, J.
1992 *Algebraic geometry: a first course*, Graduate Texts in Mathematics 133, Springer-Verlag, New York.

Hartshorne, R.
1977 *Algebraic geometry*, Graduate Texts in Mathematics 52, Springer-Verlag, New York-Heidelberg-Berlin.

Iversen, B.
1986 *Cohomology of sheaves*, Springer-Verlag, Heidelberg.

Kashiwara, M.
1976 B-functions and holonomic systems: rationality of roots of b-functions, *Invent. Math.*, **38**, 33-53.

Keller, O.H.
1939 Ganze Cremona-Transformationen, *Monatsh. für Math. und Phys.* (Leipzig und Wien), **47**, 299-306.

Kirwan, F.
1988 *An introduction to intersection homology theory*, Pitman Research Notes in Mathematics 187, Longman Scientific and Technical, Harlow, Essex.

Krause, G. and Lenagan, T.H.
1985 *Growth of algebras and Gelfand-Kirillov dimension*, Research Notes in Mathematics 116, Pitman, London.

Littlewood, D.E.
1933 On the classification of algebras, *Proc. London Math. Soc.*, **35**, 200-240.

Lunts, V.
1989 Algebraic varieties preserved by generic flows, *Duke Math. J.*, **58**, 531-554.

Malgrange, B.
1976 Le polynôme de Bernstein d'une singularité isolée, *Lecture Notes in Math. 459*, Springer-Verlag, 98-119.

1991 *Équations différentielles à coefficients polynomiaux*, Progress in Mathematics 96, Birkhäuser, Boston, Mass.-Basel-Berlin.

Manin, Yu. I.
1988 *Quantum groups and non-commutative geometry*, Publications du Centre de Recherches Mathématiques, Montréal.

Markus, L. and Yamabe, H.
1960 Global stability criteria for differential systems, *Osaka Math. J.*, **12**, 305-317.

Matsumura, H.
1986 *Commutative ring theory* , Cambridge Studies in Advanced mathematics 8, Cambridge University Press, Cambridge.

McConnell, J.C. and Robson, J.C.
1987 *Noncommutative noetherian rings*, Wiley Series in Pure and Applied Mathematics, John Wiley and Sons, Chichester-New York-Brisbane-Toronto-Singapore.

Meisters G. and Olech, C.
1988 Solution of the global asymptotic stability Jacobian conjecture for the polynomial case, *Analyse Mathématique et Applications*, Gauthier-Villars, Paris.

Milicic, D.
1986 Lectures on algebraic theory of D-modules, University of Utah.

Olech, C.
1963 On the global stability of an autonomous system on the plane, *Contributions to Differential equations*, 1, 389-400.

Pham, F.
1979 *Singularités des systèmes différentiels de Gauss-Manin*, Progress in Mathematics 2, Birkhäuser, Boston.

Pinchuk, S.
1994 A counterexample to the strong real jacobian conjecture, *Math. Z.*, **217**, 1-4.

Reiten, I.
1985 An introduction to the representation theory of Artin algebras, *Bull. London Math. Soc.*, **17**, 209-233.

Rudin, W.
1991 *Functional analysis*, second edition, International Series in Pure and Applied Mathematics, McGraw-Hill International Edition, Singapore.

Schapira, P.
1985 *Microdifferential systems in the complex domain*, Grundlehren der mathematischen Wissenschaften 269, Springer-Verlag, Berlin-Heidelberg.

Segal, I.
1968 Quantized differential forms, *Topology*, **7**, 147-171.

Serre, J.-P.
1955 Faisceaux algébriques cohérents, *Ann. of Math.*, **61**, 197-278.

Smith, S.P.
1986 Differential operators on $\mathbb{A}^1$ and $\mathbb{P}^1$ in *charp* > 0, in *Séminaire Dubreil-Malliavin 1985*, ed. M.- P. Malliavin, Lecture Notes in Mathematics 1220, Springer-Verlag, Berlin- New York, 157-177.

Stafford, J.T.
1978 Module structure of Weyl algebras, *J. London Math. Soc.*, **18**, 429-442.

1985 Non-holonomic modules over Weyl algebras and enveloping algebras, *Invent. Math.*, **79**, 619-638.

Takayama, N.
1992 An approach to the zero recognition problem by Burchberger algorithm, *J. Symbolic Computation*, **14**, 265-282.

van den Essen, A.
1991 A note on Meisters and Olech's proof of the global asymptotic stability jacobian conjecture, *Pacific J. Math.*, **151**, 351-356.

van der Poorten, A.
1979 A proof that Euler missed...Apéry's proof of the irrationality of $\zeta(3)$, *Math. Intelligencer*, **1**, 195-203.

Weyl, H.
1950 The theory of groups and quantum mechanics, Dover, New York.

Wright, D.
1981 On the Jacobian Conjecture, *Illinois J. Math.*, **25**, 423-440.

Yano, T.
1978 On the theory of *b*-functions, *Publ. RIMS, Kyoto Univ.*, **14**, 111-202.

Zeilberger, D.
1990 A holonomic systems approach to special functions identities, *J. Comp. App. Mathematics*, **32**, 321-368.

INDEX

abelian category, 166

algebraic analysis, 4

algebraic function, 41

artinian

– module, 88

– ring, 89

Artin-Schreier theorem, 89

ascending chain condition, 66

associated graded

– algebra, 57

– module, 60

asymptotically stable point, 171

asympototic stability, 171

automorphisms of A_n, 11

balanced map, 111

Bernstein

– filtration, 56

– inequality, 83

– polynomial, 94

bilinear map, 111

bimodule, 108

canonical

– basis, 9

– form, 9

category, 166

C^∞-function, 42, 45

change of rings, 130

characteristic ideal, 98

characteristic variety, 98

closed ideal, 102

co-isotropic subspace, 101

commutator, 9

comorphism, 27

complex neighbourhood, 51

composition

– of direct images, 165

– of inverse images, 140

– series, 89

degree of operator, 14

derivation, 20

difference function, 74

dimension

– of Gelfand-Kirillov, 80

– of A_n-module, 77

– of variety, 99

Dirac's δ, 46

– as hyperfunction, 51

– as microfunction, 49

direct image

– as functor, 167

– of left modules, 151

– of right modules, 146

– under embeddings, 154

– under isomorphisms, 152

– under projections, 151

direct limit, 47

directed set, 46

Printed in the United States
By Bookmasters